Advances in 21st Century Human Settlements

Series Editor

Bharat Dahiya, School of Global Studies, Thammasat University, Bangkok, Thailand

Editorial Board

Andrew Kirby, Arizona State University, Tempe, USA

Erhard Friedberg, Sciences Po-Paris, Paris, France

Rana P. B. Singh, Banaras Hindu University, Varanasi, India

Kongjian Yu, Peking University, Beijing, China

Mohamed El Sioufi, Monash University, Clayton, Kenya

Tim Campbell, Woodrow Wilson Center, Washington, USA

Yoshitsugu Hayashi, Chubu University, Kasugai, Japan

Xuemei Bai, Australian National University, Canberra, Australia

Dagmar Haase, Humboldt University, Berlin, Germany

Ben C. Arimah, United Nations Human Settlements Programme, Nairobi, Kenya

AF800557

Indexed by SCOPUS

This Series focuses on the entire spectrum of human settlements – from rural to urban, in different regions of the world, with questions such as: What factors cause and guide the process of change in human settlements from rural to urban in character, from hamlets and villages to towns, cities and megacities? Is this process different across time and space, how and why? Is there a future for rural life? Is it possible or not to have industrial development in rural settlements, and how? Why does 'urban shrinkage' occur? Are the rural areas urbanizing or is that urban areas are undergoing 'ruralisation' (in form of underserviced slums)? What are the challenges faced by 'mega urban regions', and how they can be/are being addressed? What drives economic dynamism in human settlements? Is the urban-based economic growth paradigm the only answer to the quest for sustainable development, or is there an urgent need to balance between economic growth on one hand and ecosystem restoration and conservation on the other – for the future sustainability of human habitats? How and what new technology is helping to achieve sustainable development in human settlements? What sort of changes in the current planning, management and governance of human settlements are needed to face the changing environment including the climate and increasing disaster risks? What is the uniqueness of the new 'socio-cultural spaces' that emerge in human settlements, and how they change over time? As rural settlements become urban, are the new 'urban spaces' resulting in the loss of rural life and 'socio-cultural spaces'? What is leading the preservation of rural 'socio-cultural spaces' within the urbanizing world, and how? What is the emerging nature of the rural-urban interface, and what factors influence it? What are the emerging perspectives that help understand the human-environment-culture complex through the study of human settlements and the related ecosystems, and how do they transform our understanding of cultural landscapes and 'waterscapes' in the 21st Century? What else is and/or likely to be new vis-à-vis human settlements – now and in the future? The Series, therefore, welcomes contributions with fresh cognitive perspectives to understand the new and emerging realities of the 21st Century human settlements. Such perspectives will include a multidisciplinary analysis, constituting of the demographic, spatio-economic, environmental, technological, and planning, management and governance lenses.

If you are interested in submitting a proposal for this series, please contact the Series Editor, or the Publishing Editor:

Bharat Dahiya (bharatdahiya@gmail.com) or

Loyola D'Silva (loyola.dsilva@springer.com)

Mahak Agrawal
Editor

Underground Spaces for Climate Resilience and Sustainability

Springer

Editor
Mahak Agrawal
ITACUS
New Delhi, Delhi, India

ISSN 2198-2546 ISSN 2198-2554 (electronic)
Advances in 21st Century Human Settlements
ISBN 978-981-96-0546-0 ISBN 978-981-96-0547-7 (eBook)
https://doi.org/10.1007/978-981-96-0547-7

© The Editor(s) (if applicable) and The Author(s), under exclusive license to Springer Nature Singapore Pte Ltd. 2025

This work is subject to copyright. All rights are solely and exclusively licensed by the Publisher, whether the whole or part of the material is concerned, specifically the rights of translation, reprinting, reuse of illustrations, recitation, broadcasting, reproduction on microfilms or in any other physical way, and transmission or information storage and retrieval, electronic adaptation, computer software, or by similar or dissimilar methodology now known or hereafter developed.
The use of general descriptive names, registered names, trademarks, service marks, etc. in this publication does not imply, even in the absence of a specific statement, that such names are exempt from the relevant protective laws and regulations and therefore free for general use.
The publisher, the authors and the editors are safe to assume that the advice and information in this book are believed to be true and accurate at the date of publication. Neither the publisher nor the authors or the editors give a warranty, expressed or implied, with respect to the material contained herein or for any errors or omissions that may have been made. The publisher remains neutral with regard to jurisdictional claims in published maps and institutional affiliations.

This Springer imprint is published by the registered company Springer Nature Singapore Pte Ltd.
The registered company address is: 152 Beach Road, #21-01/04 Gateway East, Singapore 189721, Singapore

If disposing of this product, please recycle the paper.

Foreword

By Han Admiraal and Antonia Cornaro, authors of "Underground Spaces Unveiled: Planning and Creating the Cities of the Future", ICE Publishing, ISBN: 978-0-7277-6145-3.

The use of underground spaces can contribute to making our cities liveable, sustainable, and inclusive. As such, we believe that the real reason why this urban frontier should be explored lies in the need to choose what kind of cities we want to live in.

We are delighted to see this book's publication on this important, constantly evolving, and emerging topic. This book unveils the immense, yet often overlooked, potential of underground spaces as crucial assets in the quest for climate resilience and sustainable urban development.

Through a rich tapestry of diverse perspectives, innovative methodologies, and illuminating case studies, it delves into how subterranean environments can offer transformative solutions to the pressing challenges posed by a changing climate and rapidly growing urban populations.

Underground spaces hold significant promise for enhancing climate resilience and promoting sustainability in urban environments. As cities grapple with the impacts of climate change and increasing populations, the innovative use of subterranean environments offers a multifaceted solution.

Underground spaces can mitigate urban heat island effects. By relocating infrastructure and activities below ground, surface areas become available for green spaces and vegetation, which naturally cool the environment. For example, underground transit systems reduce the need for above-ground roadways and parking lots, enabling more urban greenery and pedestrian-friendly spaces.

Moreover, underground spaces offer robust protection and natural insulation against extreme weather events, rendering them ideal for data centres and storage facilities that require stable conditions. This insulation also extends to residential and commercial spaces, reducing the energy required for heating and cooling, thereby lowering carbon footprints.

Therefore, this book arrives at a critical juncture, offering timely insights as we grapple with climate-related challenges that increasingly impact our daily lives.

This book titled *Underground Spaces for Climate Resilience and Sustainability, and published by Springer Nature as part of the Advances in the 21st Century series* highlights the role of underground transit systems in promoting walkable urbanism, using the Warsaw Metro System as a case study to spotlight good practices and barriers. It also explores the integration of underground archaeological spaces in downtown Amman and how such integrations can contribute to climate adaptation and sustainability while preserving cultural heritage and addressing modern urban challenges.

The book delves into the revival of both traditional and modern underground practices to mitigate heat stress in urban environments. A notable example is the ancient Qanat system in Yazd, Iran, which is examined in-depth to highlight its effectiveness in enhancing urban climate resilience and its potential applicability in contemporary settings.

Water management stands out as a critical domain where underground solutions excel. Ancient practices, such as the Qanat system, showcase how subterranean channels can efficiently manage water resources in arid climates, offering valuable insights for modern adaptations. Moreover, the book explores how underground reservoirs and drainage systems can help cities manage stormwater, reduce flooding risks, and ensure a reliable water supply during droughts.

Additionally, the book introduces the concept of geological empathy, advocating for a deeper understanding and appreciation of geological features in urban planning. This innovative approach encourages urban developers and planners to adopt a more holistic perspective, integrating the geological context into the development of urban areas. By fostering geological empathy, the book aims to inspire more sustainable and resilient urban planning practices that harmonise with the natural landscape.

The book also puts forward innovative design approaches for subway stations, aiming to transcend conventional methods by emphasising user-centric design and seamless aesthetic integration with urban environments. These proposals seek to create more functional, visually appealing, and user-friendly subway stations that enhance the overall urban experience.

In addition, the book examines the potential of subterranean spaces for geothermal energy extraction, highlighting their suitability as a sustainable and renewable energy source for urban settings. By utilising underground spaces, geothermal systems can harness the earth's natural heat for power and heating, offering a clean energy solution that reduces reliance on fossil fuels. This exploration underscores the importance of underground spaces as a viable component in the pursuit of emission reduction, carbon neutral or net-zero goals.

Importantly, *Underground Spaces for Climate Resilience and Sustainability* spotlights the role of soil life and biodiversity in underground environments, emphasising their importance as critical components of urban climate resilience. Maintaining healthy soil ecosystems supports urban infrastructure and contributes to climate adaptation efforts. Underground spaces support biodiversity and soil health, essential for a resilient urban ecosystem. Healthy soil ecosystems, maintained through strategic underground development, enhance the stability and fertility of urban landscapes.

The book concludes by summarising key insights and proposing a roadmap for future research and development in the field of underground spaces. It stresses the importance of interdisciplinary collaboration and innovative thinking in leveraging subterranean environments for urban resilience and sustainable development.

In summary, *Underground Spaces for Climate Resilience and Sustainability*, through its authors and editor, illustrates how underground spaces offer innovative solutions for climate resilience and sustainability by mitigating heat islands, protecting against extreme weather, managing water resources, supporting renewable energy, and enhancing urban biodiversity. These benefits collectively contribute to more resilient, sustainable, and livable cities. This remarkable book showcases the rapidly evolving thinking in this field and its application on the ground, across the world.

Antonia Cornaro
Chair, ITA Committee on Underground Spaces
Expert on Underground Space at AMBERG Engineering
Zurich, Switzerland

Han Admiraal
Advisor, ITA Committee on Underground Spaces
Managing Director, Enprodes Management Consultancy B.V.
Vlaardingen, The Netherlands

Contents

Beneath the Surface—Unveiling the Untapped Potential of Underground Spaces

Mahak Agrawal

Keywords Underground space · Urban underground planning · sustainability · City climate resilience · Urban innovation

Welcome to a journey into the depths—literally. As urbanisation spreads its tendrils across the globe, our cities are running out of space, upwards, outwards, and sideways. But what if we've been overlooking a dimension that holds the key to a more sustainable, resilient, and innovative future? Welcome to the underground, where the possibilities are as vast as the caverns beneath our feet.

1 The Underworld of Urbanism: More Than just a Basement

The notion of underground spaces often conjures images of dark, damp basements or ominous tunnels where secrets are buried (both literally and figuratively). But as you'll discover in the pages ahead, these subterranean spaces are far from the gloomy dungeons of our imagination. They're vibrant, dynamic, and surprisingly versatile—offering solutions to some of the most pressing challenges of our time.

In the rapidly urbanising world, the surface isn't just crowded—it's maxed out. We've built up, out, and around, but the demands on our cities keep increasing. Meanwhile, the underground remains an underutilised resource, rich with the potential to support urban expansion without the environmental costs associated with surface

M. Agrawal (✉)
ITA Committee on Underground Spaces (ITACUS), Geneva, Switzerland
e-mail: ma4100@columbia.edu

© The Author(s), under exclusive license to Springer Nature Singapore Pte Ltd. 2025
M. Agrawal (ed.), *Underground Spaces for Climate Resilience and Sustainability*, Advances in 21st Century Human Settlements,
https://doi.org/10.1007/978-981-96-0547-7_1

development. Imagine cities where traffic congestion is alleviated by underground highways, where green spaces are preserved by shifting infrastructure below ground, and where urban dwellers can escape the heat in subterranean parks.

This book dives into these possibilities, exploring how we can rethink the underground not as a last resort, but as a first choice in urban planning. By reimagining these spaces, we can create cities that are not only more functional but also more liveable, sustainable, and resilient. In the chapters that follow, we'll journey through examples from around the world, showing how different cities have tapped into the underground's potential in unique and innovative ways.

2 Digging for Carbon: The Role of Underground Spaces in Carbon Management

Let's start with one of the most pressing challenges of our time: carbon management. While the atmosphere above us is overflowing with excess carbon, the ground beneath us is quietly offering a solution. Underground spaces can play a crucial role in carbon sequestration, storing captured carbon dioxide in geological formations and turning our cities into carbon sinks rather than carbon sources.

Imagine a future where carbon is captured from the air and stored deep within the earth, safely and permanently. This isn't science fiction—it's a reality already being explored in projects around the world. Geological carbon sequestration involves injecting captured CO_2 into porous rock formations, where it's trapped by natural processes. By leveraging this approach, cities can significantly reduce their carbon footprints and contribute to global climate goals (see Fig. 1).

Fig. 1 Re-imagining underground spaces as a solution for storing carbon in geological formations and green infrastructure. Image by Mahak Agrawal

But the potential doesn't stop at storage. Underground spaces also offer opportunities for green infrastructure that can support carbon management in more subtle ways. Take biodiverse soils, for example, which can absorb and store carbon naturally while supporting urban agriculture and green spaces above. Or consider geothermal energy systems, which use the stable temperatures below ground to provide low-carbon heating and cooling solutions, reducing reliance on fossil fuels.

By thinking vertically—both above and below ground—we can develop integrated carbon management strategies that help cities contribute to climate solutions without sacrificing space or efficiency. The chapters ahead will delve into specific examples, exploring how different cities are leveraging their underground resources to tackle the carbon challenge.

3 A Fair Deal for All: Underground Spaces and the just Transition

As we transition towards a low-carbon future, it's essential to ensure that this shift is equitable—a just transition, as it's known in policy circles. Underground spaces offer unique opportunities in this regard. They can be repurposed to provide affordable housing, community spaces, and even underground urban farms that contribute to food security. These spaces can be designed with inclusivity in mind, ensuring that everyone benefits from the transition to a greener economy.

In the context of a just transition, the question isn't just how we reduce carbon, but how we do so in a way that lifts the most vulnerable among us. As cities grapple with the need to decarbonise, underground spaces offer a way to ensure that the benefits of this transition are shared by all. For instance, repurposing disused underground spaces for affordable housing can help address the urban housing crisis while reducing the environmental impact of new construction (see Fig. 2).

Moreover, underground urban farms can provide fresh produce in densely populated areas, reducing food deserts and promoting local food production. These spaces can be integrated into urban planning in a way that prioritises social equity, ensuring that marginalised communities have access to the benefits of sustainable development. It's not just about going green—it's about going green in a way that's fair and inclusive.

The chapters in this section will explore how cities around the world are using their underground resources to advance a just transition, creating spaces that are not only sustainable but also socially equitable. From underground housing projects to community gardens beneath the city streets, these examples show how the underground can be a cornerstone of a fairer, greener future.

Fig. 2 Re-imagining underground spaces' use for affordable housing and urban farming. Image by Mahak Agrawal

4 War and Peace: Resilience in the Depths

It's often said that when the going gets tough, the tough go underground. Throughout history, underground spaces have provided refuge during times of conflict, from ancient catacombs to modern bunkers. But beyond offering protection, these spaces can be designed to enhance urban resilience during warfare and other crises.

In a world increasingly defined by uncertainty—whether from geopolitical conflicts, natural disasters, or pandemics—resilience has become a key consideration in urban planning. Underground spaces offer unique advantages in this regard. They can house critical infrastructure that's protected from surface-level disruptions, provide secure transit routes in times of crisis, and serve as shelters for civilians during emergencies.

For example, underground metro systems can keep cities moving even when surface transport is disrupted. Subterranean data centres can protect valuable information from physical attacks or natural disasters. Underground bunkers, while traditionally associated with military use, can be repurposed for civilian protection, offering secure spaces for shelter during times of crisis.

But resilience isn't just about defence—it's about adaptability. Underground spaces can also be designed to be multifunctional, serving different purposes in different contexts. For instance, a space that serves as a parking garage during peacetime could be converted into a shelter or command centre during emergencies. By building flexibility into these spaces, cities can enhance their resilience to a wide range of challenges, both anticipated and unforeseen (see Fig. 3).

The chapters on this theme examine how cities around the world are using underground spaces to enhance their resilience, drawing on lessons from history as well as cutting-edge innovations. Whether it's protecting infrastructure, providing shelter,

Fig. 3 Underground metro stations for shelter. Image by Mahak Agrawal

or maintaining essential services, these examples highlight the critical role of the underground in ensuring urban resilience.

5 Art Underground: Creativity as a Catalyst for Innovation

Before you start thinking that underground spaces are all about doom and gloom, let's take a moment to explore how art can breathe life into these subterranean environments. Underground art installations, like New York City's vibrant subway murals or Stockholm's stunning tunnelbana, prove that creativity doesn't stop at street level. In fact, it can flourish below ground, transforming utilitarian spaces into inspiring environments.

Art has always been a powerful force for reimagining spaces, challenging us to see the world differently. When it comes to underground spaces, art can play a transformative role, turning functional, often overlooked environments into spaces of wonder and inspiration. From murals that brighten the daily commute to sculptures that bring a sense of whimsy to otherwise stark environments, art can make the underground not just bearable, but enjoyable (see Fig. 4).

Moreover, art can drive innovation in how we design and use underground spaces. By challenging conventional notions of what these spaces can be, artists inspire us to think outside the box—or in this case, outside the bunker. For example, art installations can help raise awareness about the importance of underground spaces for climate resilience or carbon management, engaging the public in conversations that might otherwise seem abstract or technical.

But the value of art underground goes beyond aesthetics. It can also foster a sense of community, bringing people together in spaces that might otherwise feel isolating

Fig. 4 Underground space used for art installations and creative display. Image by Mahak Agrawal

or impersonal. Imagine an underground gallery where local artists showcase their work or a subterranean park where public art invites interaction and engagement. These spaces can become cultural hubs, enriching the urban fabric and making our cities more vibrant and inclusive.

The chapters on this theme will showcase examples of how art is being used to reimagine underground spaces, from public installations to innovative design approaches. These stories illustrate how creativity can turn even the most utilitarian spaces into places of beauty and connection, proving that the underground is anything but dull.

6 The Road (or Tunnel) Ahead

As you turn the pages of this book, prepare to have your preconceived notions of underground spaces turned on your head. You'll see how these spaces can contribute to a just, resilient, and sustainable future—whether through carbon management, providing shelter during crises, or serving as canvases for artistic expression.

So, grab your hard hat and your curiosity, because we're about to embark on a journey into the underground. It's time to stop overlooking what lies beneath our cities and start embracing the untapped potential of the world beneath our feet. Who knows? By the end of this book, you might just find yourself looking down in admiration at the ground you walk on, knowing that the future of urban innovation lies below.

Beneath the Surface: Foundations of Underground Urbanism

Do Look Down: Reviving Underground Urban Spaces to Adapt to Heat Stress

Tjark Gall

Abstract Climate change is increasing pressure on cities: high temperatures, intensified through the heat island effect, are increasing health concerns, and energy demand while reducing productivity. Common solutions include air conditioning or improving insulation—while ignoring the near infinite amount of cooling potential that lies just beneath our feet. Only metres below the surface, the subsurface temperature can significantly improve thermal comfort during heatwaves. Depending on geology, depth, season, and climate, the underground provides a reliable cooling source. In response to the climate awareness-building movie 'Don't Look Up', this contribution invites the reader to look down instead. Building on past urban underground spaces (UUS) and a literature review on using UUS to adapt to climate change, this chapter advocates for considering UUS in the battle against urban heat. It compiles possible functions of UUS and key considerations to respond to the challenge of maintaining *public* space. By combining two possible climate change pathways and varying levels of UUS use scenarios, the chapter highlights four scenarios to showcase the impact potential of UUS' in adapting to climate change and mitigating the urban heat island effect.

1 Introduction

In the face of ongoing rapid urbanisation and extreme heat challenges, there is a growing need to rethink contemporary options of urban living. This term encompasses the shared utilisation of public spaces, which is becoming increasingly challenging globally due to rising heat stress levels.

This chapter presents a creative yet scientifically grounded exploration of leveraging underground spaces to create innovative, liveable urban environments inspired by historic practices and adapted to contemporary needs. Building upon the historical precedent of underground uses, such as subterranean dwellings in early settlements in

T. Gall (✉)
Washington, D.C., United States of America
e-mail: tjark.gall@urban-framework.com

© The Author(s), under exclusive license to Springer Nature Singapore Pte Ltd. 2025
M. Agrawal (ed.), *Underground Spaces for Climate Resilience and Sustainability*, Advances in 21st Century Human Settlements,
https://doi.org/10.1007/978-981-96-0547-7_2

hot climates, this contribution integrates social, architectural, and urban perspectives to propose novel approaches for underground living in modern cities.

Aside from potentially reviving, scaling, and innovating public uses of urban living, such as heat shelters, storage facilities with reduced cooling demand, or sports facilities, the contribution explores the role of recent technological advances. Some of the most common concerns of underground uses are high costs for excavation and stabilisation works and lack of natural daylight, fresh air, and liveable humidity levels. However, combining traditional practices with nature-based solutions and technological innovation—both high- and low-tech—provides ample opportunities.

Central to this work is recognising the inherent potential of underground spaces as a sustainable solution to urban heat stress. By tapping into the thermal stability of the earth and leveraging historical practices, this contribution seeks to foster new paradigms of urban design and architecture that prioritise environmental resilience and human well-being by considering a different perspective than the usual upward one.

The contribution provides a short overview of historic uses across cultures and contemporary innovative solutions that inform four scenarios of urban underground uses of tomorrow—while combining a descriptive academic perspective with a prospective creative lens.

This work uses previously developed prospective methods on future scenarios and builds on past and ongoing works on climate change adaptation and mitigating urban heat in particular. While discussions on thermal dynamics and comfort span both cold and warm, the focus lies on mitigating urban heat. While cold-related deaths outnumber heat-related ones, heat stress is rising due to climate change, dominating in low-resource areas, and historically less addressed [1]. As urban heat is an issue that affects most cities in the world, this contribution does not limit itself to one area or the other. However, the prospective part targets megapolitan[1] urban areas in more extreme climates as well as a more moderate European example.

2 Escaping Urban Heat

More people live on the planet than ever before, increasingly more in urban areas, especially metro- or megapolitan areas in the global south. Combined with climate change and increase in temperature as well as intensity and frequency of disasters, adapting to urban heat is one of the pressing challenges of this century [2, 3].

Increasingly, cities—even northern European ones—develop urban heat adaptation strategies, emergency protocols, and hire chief urban heat officers to raise awareness and mainstream adaptation actions. The reasons are simple: Rising excess

[1] Defined as a continuous, historically, and culturally homogenous region that combines existing metropolitan areas, houses more than 10 million people, and is connected via major transportation infrastructure. (Source: https://www.lincolninst.edu/publications/articles/americas-megapolitan-areas).

mortality, other health issues, increasing energy demand for cooling through air conditioning (AC), loss of economic productivity, and a highly unequal distribution of risk and impact on vulnerable groups, especially the elderly and low-income [3].

Aside from the concentration of people in urban areas, the urban heat island (UHI) significantly aggravates the situation (Fig. 1). UHI effect refers to the phenomenon where urban areas experience higher temperatures compared to their surrounding rural areas, primarily caused by the absorption of more solar radiation by urban infrastructure, reduced coverage of naturally cooling vegetation and water bodies, higher usage of heat-absorbing building materials, reduced air circulation, and higher output of anthropogenic heat sources such as waste heat from vehicles and ACs [4]. The UHI effect increases energy needs and potential public health crises further while continuously reducing labour productivity in urban areas [5].

The part of the population most affected by urban heat and the UHI effect are those living in urban areas, particularly in the global south and South Asia. South Asia, home to a quarter of the world's population, is highly vulnerable to the impacts of urban heat, and the number of people impacted by warming temperatures is set to rise, with two out of three people living in an urbanised area by 2050 [4]. Estimations link several thousand heat-related deaths in India and Pakistan to heat waves [3]. The UHI effect is expected to increase due to the projected rise in global average temperatures caused by climate change [2]. Global surface temperatures have already risen 1.1 °C above pre-industrial levels and are continuing to rise, leading to an increase in the frequency, severity, and intensity of heat waves [4]. By 2100, 25% of the world's largest cities could warm by up to 7 °C, and the number of people impacted by this is set to rise with 70% of the global population projected to live in cities by 2050 [2, 4, 6].

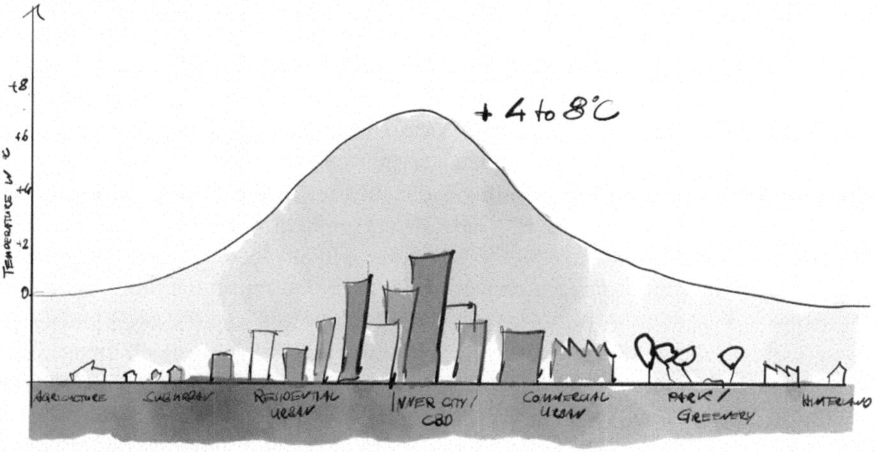

Fig. 1 Diagram showing conceptual differences in temperature due to the heat island effect (the author)

Aside from the urgent need to reduce the temperature increase by slowing down or reversing the climate crisis and adapting to the heat through local solutions—such as passive cooling, increasing water features, and nature-based solutions. However, underground spaces as a solution space to urban heat stress have gained little attention.

2.1 *Underground Spaces as Adaptation Strategy*

The concept of underground urbanism and associated works are on the rise. However, targeted research on cooling through underground spaces and climate adaptation in general are still limited. A search on the Web of Science of related terms[2] led to 209 initial results, reduced to only five following a title-and abstract-based preselection. These five articles were complemented by two referenced articles. The following section provides a short summary of the current state of the art on urban underground spaces (UUS) for climate adaptation and heat in particular.

Admiraal and Cornaro—frontrunners of underground urbanism—argue that UUS provide 'ecosystem services to the city and offers invaluable resources in the form and shape of space, water, energy, and minerals' but that an integrated approach is so far missing [7]. They further refer to a UN document that highlights that UUS can serve multiple disaster risk reduction functions such as temporary water retention basins or refuges. The potential of double use of UUS is emphasised, e.g., to store water during floods and use it most of the time as parking, also to recover costs [7]. Other climate adaptation functions mentioned are the provision of shelter and storage space, heating or *cooling, and protection from earthquakes* [8]. Others join in highlighting the role of UUS to reduce high outdoor temperatures in hot climates, with sunken courtyards in Iran serving as an example [8]. Bobylev et al. call for better integration of UUS in the design process of climate-responsive cities, providing the example of the Paris metro system that was used as shelter for the elderly during summer heat waves [8].

Tan et al. studied the rising exploitation of underground spaces in China, planned through the national 'Plan for Urban Underground Space Planning' that calls for a comprehensive system for UUS planning and management in half of Chinese cities by 2020 [9]. The authors analysed over 2,000 underground projects built between 2011 and 2015, often contributing to eco-friendly developments by reducing their surface-level impact. Functions span metro stations/complexes, museums, archaeological tomb palaces, basement dormitories, civil shelters, power/water/waste plants, and commercial basements [9]. However, they also mention some of its challenges: (i) The negative perception of the general population, linking UUS to environmental stressors and negative psychological effects (cf. Roberts et al. 2016 in [9]), (ii) users expressing their concern about insufficient ventilation, the importance of effective evacuation and maintenance of standard operation procedures and regular emergency

[2] Search query: (*city* (Abstract) and *underground* (Abstract) and *heat* (Abstract) or *adaptation* (Author)), last updated on April 26, 2024.

drills, as well as the possible association of UUS with negative energy in Feng-Shui theory [9]. On the positive side, Tan et al. highlight that working in UUS can improve the thermal comfort of workers during cold periods—something that should hold true for heat waves. They highlight that for effective human use of UUS, several key factors have been identified: (i) Good indoor quality, (ii) velocity of air/wind impacting comfort/perceived temperature, (iii) good acoustics, and (iv) greenery that mitigates the feeling of being inside [9].

Zhou et al. analyse the building and architectural quality dimension, focusing on the potentially increased light and ventilation energy consumption [10]. On the other hand, they highlight the diverse and increasing number of solutions, e.g., passive lighting, wind- or buoyancy-driven ventilation, sunken plazas, skylights, and light wells. To ensure and measure the level of thermal and overall perceived comfort, they developed and adapted many metrics, including natural lighting comfort, natural ventilation performance, and thermal comfort [10].

Zooming out from the architectural scale, a correlation exists between population and building density and the exploitation of UUS: Across global cities, Chen et al. revealed that high density 'is the driving force of urban underground', in Chinese cities especially dominated by residential and commercial underground parking and commercial activities, with some limited functions of culture, medical, sports, education, entertainment, and business [11].

This aggregation of surface and underground functions and infrastructures in a limited area—fuelled by peak space demand resulting from centrality and accessibility[3]—requires integrated planning in harmony with other urban systems. Finesso and Van Ree claim that UUS are in urban planning processes 'virtually absent, rendering it an underappreciated asset' [12]. Its regulation and potential planning are complex due to the overlap of various 'geosystem services'. The concept distinguished between (i) provisioning services, (ii) cultural services, and (iii) regulating services [12]. The latter is of particular relevance to urban heat adaptation.

Von der Tann et al. focus further on the planning and regulation of UUS and highlight the strong variety of how (much) they are regulated [13]. The authors reviewed several national planning policy guidelines that contain relevant topics for UUS and identified the following twelve: (i) Air quality, (ii) Climate change, (iii) Conserving and enhancing the historic environment, (iv) Environmental impact assessment, (v) Flood risk and coastal change, (vi) Land affected by contamination, (vii) Land stability, (viii) Minerals, (ix) Natural environment, (x) Water, and (xi) Water supply, wastewater, and water quality [13].

In summary, limited dedicated literature exists on using UUS for climate change adaptation. However, the presented literature provides a useful framework on what aspects are important to consider when exploring the utility of UUS for urban heat stress mitigation.

[3] See, for example, Alain Bertaud's 'Order without Design' for a detailed account on land demand dynamics and their impact on urban form, the distribution of functions, and prices.

2.2 Examples of Urban Underground Spaces

While little discussion can be found in current literature, the concept of using UUS is not new. In this section, a few historic and more recent examples are presented to feed into their potential renaissance at a larger scale.

One of the most noteworthy examples is the Derinkuyu Underground City in Turkey, where recently, an ancient underground city has been discovered (Fig. 2). Derinkuyu is up to 18 stories (76 m) deep and has over 15,000 shafts for ventilation [14]. The higher and better ventilated parts were most likely used as living and sleeping areas while the lower ones were primarily for storage [14]. A quoted theory states that Derinkuyu 'served as a temperate refuge for the region's extreme seasons' with Cappadocian (its region) winters becoming very cold and summers extremely hot [14].

Derinkuyu is just one—and with up to 20,000 inhabitants in the past possibly the largest—of up to 36 cities in the Cappadocia region [15].

UUS are constrained by various factors, especially geological conditions, excavation technology, and technical equipment (e.g., heating, electric installations, air conditioning, water, canalisation) [15]. Nevertheless, Nývlta et al. compiled a variety of dominating use cases: (i) Escape routes due to conflict, (ii) Lack of surface space, (iii) Need for protection against the cold, dust, or noise pollution, and (iv) Preservation of historical and cultural heritage. Depending on their use and location, UUS are usually made up of larger caverns for school or religious or social gatherings, bedrooms, kitchen, bathrooms, stables for animals, water tanks, water wells, food and wine storage, separate tombs and burial places, and weapons depots [15].

Fig. 2 Artistic representation of the underground city Derinkuyu (*Source* Yasir999, CC BY-SA 4.0)

Aside from Derinkuyu and similar UUS in Cappadocia, other historic and more recent examples have been compiled in Table 1. While not discussed in detail due to this chapter's scope, they highlight the diverse use of underground spaces over time and context.

A more recent example is that of Palestine, especially the Gaza Strip, and the Ukraine. The underground infrastructure has found ample attention in the recent media discourse with several tunnels being used to hide from attacks and move and supply resources in times of ongoing war and geopolitical crisis. Despite risks associated with tunnel and underground operations, they have proved crucial for

Table 1 List of underground uses in history (author's compilation)

Name	Context	Use
Qanat	Qanats, also known as underground aqueducts, have been used in Iran and other parts of the Middle East for over 3,000 years	They are underground channels that use groundwater sources to transport water over long distances to supply agricultural fields and communities. These water management systems are still in use today, play a crucial role in sustaining agriculture and human settlement in arid regions, and can function as additional cooling sources in hot areas
Cappadocia Underground Cities	Cappadocia, located in central Turkey, is home to several underground cities dating back to ancient times	The underground cities were most likely carved out for different reasons, including early Christians escaping persecution and other conflict parties
Derinkuyu Underground City	Located in Cappadocia, Turkey, Derinkuyu dates back to the Byzantine era, possibly as early as the eighth century BCE	Initially used as a shelter to protect inhabitants from invasions and natural disasters, Derinkuyu could accommodate up to 20,000 people along with their livestock
Naples Underground	The underground city of Naples (Italy) dates back to ancient Greek and Roman times but grew further during the Middle Ages and Renaissance	Initially used as quarries to extract tuff for construction, these tunnels later served as shelters during World War II bombings. Today, they are a popular tourist attraction, house art exhibitions and cultural events
Wieliczka Salt Mine	Located near Krakow, Poland, the mine has been in operation since the thirteenth century	The underground complex served as a salt mine for centuries and features intricate sculptures, chapels, and even an underground lake. It is now a UNESCO World Heritage Site and a popular tourist destination
Matmata Underground Troglodyte Houses	Located in the Matmata region of Tunisia, these underground dwellings date back centuries and were inhabited by the Berber people	The houses were carved into soft limestone to provide shelter from the desert heat, featuring interconnected rooms arranged around a central courtyard

(continued)

Table 1 (continued)

Name	Context	Use
Edinburgh Vaults	Built in the late eighteenth century under the streets of Edinburgh, Scotland	Originally planned as storage space for merchants, the vaults later became overcrowded slums and were associated with crime and poverty. Today, they are a popular tourist attraction and part of various legends
Catacombs of Paris	Created in the late eighteenth century beneath the streets of Paris, France	The catacombs were used to store the remains of millions of people transferred from overcrowded cemeteries. Today, they are another famous tourist attraction and historical and cultural landmark
Seattle Underground	Established in the late nineteenth century beneath the streets of Seattle, Washington	After a devastating fire in 1889, Seattle rebuilt its downtown area one story higher, leaving the original ground-level storefronts and sidewalks underground. The area remains accessible for tourists to visit the old passages and remnants
Beijing Underground City	Built during the Cold War era in the 1960s under the streets of Beijing, China	Constructed as bomb shelters and nuclear fallout shelters in anticipation of a potential conflict with the Soviet Union or the United States. The underground city featured living quarters, hospitals, schools, and even theatres
Coober Pedy	Located in the Australian outback, Coober Pedy was established in the early twentieth century as a mining town for opals	The extreme heat of the outback drove residents to build underground homes, known as 'dugouts', to escape the harsh climate. These underground dwellings offer natural insulation and protection from temperature extremes
Montreal Underground City	Developed in the mid-twentieth century below the streets of Montreal, Quebec	Designed to provide shelter from harsh winter weather, the Montreal Underground City 'RESO', consists of connected tunnels, shopping centres, office buildings, hotels, and metro stations
Singapore Underground	Singapore has developed an extensive underground infrastructure to maximise land use and alleviate congestion in the densely populated city-state	The Singapore Underground includes underground malls, pedestrian walkways, transportation tunnels, and utility networks. These underground spaces provide relief from the tropical heat, offer convenient connections between buildings, and ensure efficient delivery of essential services
Hong Kong Underground	Hong Kong, known for its high population density and limited land availability, has an extensive underground network	The Hong Kong UUS includes shopping malls, pedestrian walkways, transportation hubs, and utility tunnels. These facilities reduce surface congestion, provide shelter during extreme weather, and support the efficient functioning of the city's infrastructure

maintaining economic activity in Gaza and protecting people in several areas of Ukraine, especially Kharkiv.

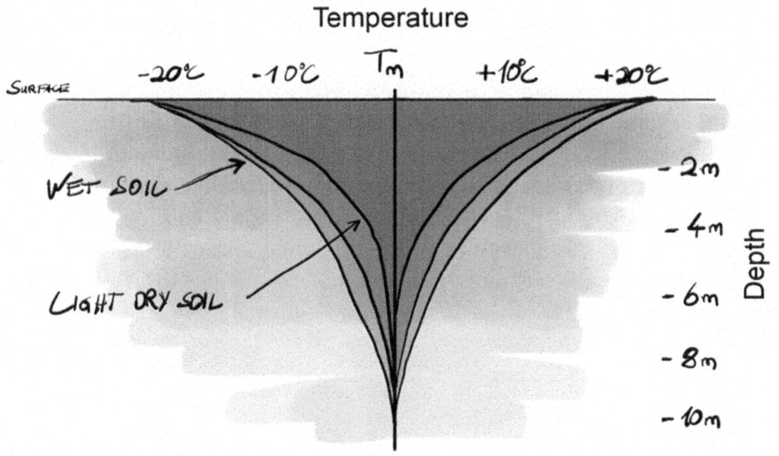

Fig. 3 Conceptual visualisation of decreasing temperatures by depth (adapted from [17])

2.3 Linking Underground and Temperatures

As mentioned in several examples above, the ground has a special function to moderate temperatures. The underground temperature remains more stable compared to surface temperatures. The way it changes varies depending on soil types, climate, and depth. In colder climates, the underground temperature tends to be warmer than the surface during winter due to geothermal heat. Conversely, in warmer climates, the underground temperature may be cooler than the surface during summer. Soil types also play a role; for instance, dry, sandy soils have greater temperature fluctuations compared to moist, clay-rich soils, which tend to insulate better. Figure 3 shows a simplified visualisation of the underlying geophysical concept: Starting from a certain depth (commonly between 6 and 10 m), the soil has a relatively stable yet location- and soil-specific temperature throughout the year. The colder (left side of the graph) or hotter (right side) the surface temperature is, the deeper one must go to moderate the temperature. In other words, regardless of climate, location, and soil type, the 'right' depth can always provide the required environmental temperature; and these depths are for cooling usually not that deep.

2.4 Moving Public Space Down?

This chapter so far referred to UUS as space, focusing especially on public functions yet providing examples of private residential and commercial functions, disaster risk management, as well as lifelines for conflict regions. The different management systems of existing underground spaces, as well as the ongoing high exploitation costs of subsurface developments, result in a dominance of semi-public or private underground spaces.

Public space in the urban area however refers to areas that are accessible to everyone and designed for communal use, such as parks, plazas, streets, and squares. These spaces play a vital role in society to foster social interaction, community engagement, and cultural exchange. Lefebvre introduced the concept of 'the right to the city', emphasising the idea that urban spaces should be accessible, inclusive, and shaped by the people who inhabit them. Public spaces are not just physical environments but social constructs fundamental for people to assert their rights, identities, and aspirations [16].

3 Future Scenarios

So far, this contribution presented the increasing challenge of urban heat, past use cases of the sub-terrain for adaptation, technological innovations enabling increasing UUS exploitation, the largely underexplored cooling potential of UUS, and finally the challenging but important role of preserving key principles of public space when planning the underground spaces of tomorrow.

Here, four scenarios are developed using earlier adapted prospective methods [18] and the 2*2 scenario matrix with two *critical uncertainties*: (i) Different levels of climate change on one axis and (ii) different levels of underground exploitation (e.g., due to costs) on the other.

For the climate change level, IPCC Representative Concentration Pathways (RCPs) are used. RCPs describe different pathways of greenhouse gas emissions and atmospheric concentrations, air pollutant emissions, and land use. One intermediate (RCP4.5) and one extreme (RCP8.5) pathway are chosen. The latter signifies, for example, the concentration of carbon that delivers global warming at an average of 8.5 watts per square metre with an average global temperature increase of about 4.3 °C by 2100. The four resulting scenarios (Table 2) combine the contextual uncertainties and the previously presented considerations for UUS for climate adaptation:

Scenario 1 'Underground Cooling' combines limited climate change and reduced UUS exploitation. While underground space is not used for people, colder air is used more extensively for cooling surface infrastructure, including the use of traditional techniques such as cooling air by circulating it through small underground water canals. In case of extreme heat waves, the spaces remain accessible as shelters for people who do not have access to cool places otherwise.

Scenario 2 'Underground Plazas' is a response to a scenario with high temperatures and limited underground exploitation. This could result, for example, in plazas that combine a partially open-air square and park with water features and wind towers, providing an accessible yet colder place in the urban environment.

Table 2 Visualisations of four different future scenarios based on two uncertainties: (i) Level of underground exploitation and (ii) Climate change[a]

	Continued climate change (IPCC RCP4.5)	Accelerated climate change (IPCC RCP8.5)
Low underground exploitation		
High underground exploitation		

[a] The scenario visualisations have been created using the generative AI tool Midjourney, based on the provided scenario descriptions

Scenario 3 'Underground Places' combines high UUS exploitation with intermediate climate change and resulting increasing heat. Underground areas provide a range of new functions and provide additional, shaded public spaces and squares where people can gather or cool down.

Scenario 4 'Underground Shelter' matches accelerated climate change with extensive underground exploitation. A possible result would be the increased use of underground spaces for residential functions to increase thermal comfort and reduce energy needs, using passive natural light and air circulation, e.g., through adapted wind towers, to ensure liveable spaces.

4 Discussion and Conclusion

Temperatures will rise and require new and innovative thinking. *Looking down*, if planned and implemented well, can provide a wide opportunity space that so far is underexploited. Further, there is the risk that while prices for underground use will most likely continue to decrease, costs may increase to redirect or circumvent infrastructure installed in the meantime.

Planning urban systems in an integrated and future-ready manner is therefore a key requirement, including underground spaces. Further, when designing possible pathways to sustainable future urban systems, the adaptation potential found underground should be considered.

While this chapter only provided a snapshot, it seeks to highlight key problems and directions for future work. Aside from planning, the revival of ancient cooling techniques in combination with new passive and low-energy technologies for lighting and ventilation, are critical ingredients of increasing use of UUS. While developing underground remains more expensive than surface buildings, increasing population pressures in urban areas and improving technology are probable enablers. Further, as highlighted by Admiraal and Cornaro [7] the costs can be further brought down if underground spaces are used for multiple purposes, such as underground parking and flood water retention (cf. Rotterdam Centraal parking). If all these issues might be resolved, it is important to keep the qualities of public space in mind, as well as their users: How do underground spaces need to be designed to respond to the needs of urbanites? What can be done to ensure that people are comfortable and (feel) safe?

These are only some questions that remain to be answered. Nevertheless, next time, when the sun is burning especially strong, and you are looking for solutions for more safe, productive, and liveable cities—*Do look down instead*.

Competing Interests The author has no conflicts of interest to declare that are relevant to the content of this chapter.

References

1. Zhao Q et al (2021) Global, regional, and national burden of mortality associated with non-optimal ambient temperatures from 2000 to 2019: a three-stage modelling study. Lancet Planet Health 5:e415–e425. https://doi.org/10.1016/S2542-5196(21)00081-4
2. IPCC (2022) Climate change 2022: impacts, adaptation and vulnerability. contribution of working group II to the sixth assessment report of the intergovernmental panel on climate change [Pörtner H-O, Roberts DC, Tignor M, Poloczanska ES, Mintenbeck K, Alegría A, Craig M, Langsdorf S, Löschke S, Möller V, Okem A, Rama B (eds)]. Cambridge University Press, Cambridge and New York. https://doi.org/10.1017/9781009325844
3. Kim E, Henry G, Jain M (2023) Urban heat in South Asia: integrating people and place in adapting to rising temperatures. Policy Brief. Washington D.C. The World Bank
4. Roberts M, Chandan D, Nicholas J, Park J (2023) Unliveable: what the urban heat island effect means for East Asia's cities. Washington D.C. The World Bank

5. World Bank Group (2021) Climate change action plan 2021–2025. South Asia roadmap. Washington D.C. The World Bank
6. Estrada F, Botzen W, Tol R (2017) A global economic assessment of city policies to reduce climate change impacts. Nature Clim Change 7:403–406. https://doi.org/10.1038/nclimate3301
7. Admiraal H, Cornaro A (2018) The contribution of urban underground space to climate-proof and resilient cities. In: 54th ISOCARP congress 2018
8. Bobylev NG, Guo D, Benardos A (2023) Urban underground space use in a climate neutral city. In: Anagnostou B, Marinos (eds) Expanding underground. Knowledge and passion to make a positive impact on the world. ISBN: 978-1-003-34803-0
9. Tan Z, Robert AC, Christopoulos GI, Kwok K-W, Car J, Li X, Soh C-K (2018) Working in underground spaces: architectural parameters, perceptions and thermal comfort measurements. Tunn Undergr Space Technol 71:428–439. https://doi.org/10.1016/j.tust.2017.09.002
10. Zhou K, Leng J-W (2023) State-of-the-art research of performance-driven architectural design for low-carbon urban underground space: systematic review and proposed design strategies. Renew Sustain Energy Rev 182:113411. https://doi.org/10.1016/j.rser.2023.113411
11. Chen Y, Chen Z, Guo D, Zhao Z, Lin T, Zhang C (2022) Underground space use of urban built-up areas in the central city of Nanjing: insight based on a dynamic population distribution. Undergr Space 7:748–766. https://doi.org/10.1016/j.undsp.2021.12.006
12. Finesso A, van Ree CCDF (2022) Urban heat transition and geosystem service provision: a trade-off? A study on subsurface space scarcity in the city of Amsterdam. TunnLing Undergr Space Technol 128:104619. https://doi.org/10.1016/j.tust.2022.104619
13. von der Tann L, Metje N, Admiraal H, Collins B (2018) The hidden role of the subsurface for cities. Proc Inst Civ Eng—Civ Eng 171(6):31–37. https://doi.org/10.1680/jcien.17.00028
14. Jacobs F (23 Dec 2023) Derinkuyu: mysterious underground city in Turkey found in man's basement. Big think, strange maps. https://bigthink.com/strange-maps/derinkuyu-underground-city/. Accessed 20 May 2024
15. Nývlta V, Musílek J, Cejka J, Stopka O (2016) The study of Derinkuyu underground city in Cappadocia located in pyroclastic rock materials. Procedia Eng 161:2253–2258. https://doi.org/10.1016/j.proeng.2016.08.824
16. Lefebvre H (1974) La production de l'Espace. Éditions Anthropos, Paris
17. Built Solar (201x) Ground temperatures as a function of location, season, and depth. https://www.builditsolar.com/Projects/Cooling/EarthTemperatures.htm. Accessed 10 May 2024
18. Gall T, Vallet F, Yannou B (2022) How to visualise futures studies concepts: revision of the futures cone. Futures 143:103024. https://doi.org/10.1016/j.futures.2022.103024

Exploring Warsaw Metro: Subterranean Solutions for Walkability in Cities

Annika Lundkvist

In an era of increasing urbanization and competing interests for city spaces, the demand for innovative and practical solutions is higher than ever. Utilizing urban underground space is increasingly becoming essential where it makes geological and spatial sense. As cities globally shift toward sustainable mobility, it's crucial to reassess the disproportionate allocation of subsurface space to cars and explore ways to prioritize pedestrian movement. This shift addresses spatial equity, improves air quality, protects the environment, and enhances public health and well-being in cities and regions. Using subsurface space for mass transit supports this priority and highlights the potential for underground spaces to function as walkable layers of the city. Established in the mid-1990s, Warsaw's metro system exemplifies best practices in creating a walkable urban underground environment.

1 Introduction

A relative newcomer and modern addition to the world's subway system networks, Warsaw's current 2-line metro system embodies diverse design and accessibility features, offering an exemplary case study of walkable urbanism in underground metro systems and space. Such an example is crucial as, more than ever, prioritizing pedestrian mobility and safe, equitable access to quality public transportation systems is paramount for encouraging a societal shift to more sustainable modes of mobility, reducing car dependence and encouraging the use of public transportation.

A. Lundkvist (✉)
Institute of Geography and Spatial Organization, Warsaw, Poland
e-mail: annika@twarda.pan.pl

© The Author(s), under exclusive license to Springer Nature Singapore Pte Ltd. 2025
M. Agrawal (ed.), *Underground Spaces for Climate Resilience and Sustainability*,
Advances in 21st Century Human Settlements,
https://doi.org/10.1007/978-981-96-0547-7_3

1.1 Climate Resilience Sustainable Mobility in the Underground

Humans have been using underground space for walking and passage since ancient times. The spatial importance of the subsurface for a walkable community is imperative in many regions where extreme weather necessitates the use of underground spaces by pedestrians. Beyond extreme temperatures, scarcity of land above ground, the need for preserving and expanding green spaces, security concerns, and land prices necessitate the exploration of underground landscape.

In 2014, ACUUS (Associated Centers for the Urban Underground Space) signed an agreement with UN-Habitat to spread awareness of underground spaces' role in urban drainage, city planning, and sustainable mobility.[1]

An important first step in understanding the benefits that the urban underground can offer is raising awareness,[2] which this paper intends to highlight with a focus on the potential benefits of walkable underground spaces.

The chapter explores walkability in and around the Warsaw metro system to highlight the benefits and challenges of using underground spaces. It focuses on the behavioral and experiential aspects of pedestrians using urban underground space through interviews and a literature review. It includes insights from established researchers, practitioners in subsurface space, and representatives from Warsaw Metro, also known locally as Metro Warszawskie.

As cities shift toward more sustainable mobility, many are expanding their metro systems. The use of underground transit to alleviate traffic congestion is well-recognized and established. Mass underground transit reduces surface traffic, allowing for other uses.

While knowledge of constructing underground spaces is prevalent in urban development discourse, the focus on the user experience and human perspective is less addressed.[3] Given the high cost of developing subsurface spaces, maximizing their utility and creating well-developed, sustainable environments is crucial.

Developing these projects with a focus on best practices for underground walkability ensures these spaces are widely used and appreciated. Analyzing good examples in various systems helps to enhance human comfort, safety, and accessibility in underground environments. It is thus imperative to deploy a holistic approach to urban planning for sustainable development in the twenty-first century. Utilizing underground spaces for societal mobility is a critical component of this effort (see Fig. 1).

---

[1] UN-HABITAT and ACUUS [1].

[2] Broere [2].

[3] Besner [3].

Fig. 1 Bródno station, *Image source* Metro Warszawskie

1.2 Underground Pedestrian Mobility

Challenges in planning urban underground space are numerous and often include lack of connectivity to nearby buildings or modes of transportation [3].

The purpose of the underground pedestrian network connected with metro systems has been defined in terms of its functional utility. Such systems are designed to:

1. Link urban activities with public transportation networks.
2. Mitigate conflicts between pedestrian and automobile traffic above ground.
3. Protect from extreme weather and provide thermal comfort.
4. Maximize access to urban space above ground for food security, green infrastructure, etc.[4]

Many metro operators are not considerate of pedestrian behavior inside a corridor. Pedestrians are not only walking, but also thinking about where they are going, how they feel safe, what they are hearing, on what type of floor they are walking, or how they can go outside. For a successful pedestrian system to function and thrive, the designer needs to truly understand the system's users. It is important for the designers and architects to provide protection, comfort, and pleasure for its diverse group of pedestrians.

Pedestrians have different needs and move at different speeds, with different mobility behaviors. It is crucial to consider such differences—such as people with

[4] Zacharias [4].

reduced mobility, elderly people, shoppers of various age groups, people with children, strollers, and more—and accommodate them in the design and management of the pedestrian system.

2 Underground Pedestrian System of Warsaw Metro

Opened in 1995, the Warsaw metro (Metro Warszawskie) is currently the only subway system in Poland. It operates as a two-line system with plans for expansion to five lines. The system is operated by Metro Warszawskie sp. z o.o., a municipally owned company, and managed by ZTM (Zarząd Transportu Miejskiego w Warszawie).

The metro is regarded by its operating company as the most efficient means of transport that also meets the essential ecological parameters for the sustainable development of the city, without causing excessive ground traffic.[5] Warsaw is a rapidly expanding city, with new, highly dense housing estates being built in post-industrial areas. An area of particularly intense change is the area of Kamionek, Gocławek, and Gocław in Praga Południe. Therefore, Mr. Rafał Trzaskowski—President of the Capital City of Warsaw, decided to expand metro lines.

Urban governance is crucial for developing comprehensive systems that integrate walkable surface and subsurface layers. The Warsaw Metro aims to overcome mobility barriers across the city by creating convenient transfer hubs and prioritizing accessibility features, such as escalators and elevators, throughout the system. The focus is on providing a comfortable, accessible, and safe metro system for passengers.

Warsaw has been selected as one of 100 European pioneer cities to participate in the Mission for Climate Neutral and Smart Cities by 2030. This participation offers the capital an opportunity to develop specific solutions to key climate protection challenges. The signed agreement marks an important step in this direction. Additionally, Warsaw's collaboration with the Kingdom of Denmark, aligned with the Green Vision of Warsaw, will contribute to improving the city's air quality.

In an interview, José Antonio Juncà mentioned "If we really want to work for sustainability and accomplish the SDGs of Agenda 2030 of the United Nations, for example, Objective nº 11, efforts should be enhanced to improve the green and blue City, reducing the emissions of CO_2, calming transit, improving Universal Accessibility, working to achieve 'slow cities', less stressed urban spaces; and to reach all those goals, more clever use of underground space for public use should be made, providing innovations, and creating links between the above-ground and the underground city, in a more continuous way."[6]

Ongoing improvements to diverse aspects of the pedestrian networks in the city play a significant role in the overall climate resilience strategies of the city, easing the decision for inhabitants to choose sustainable modes over routine car dependence. Ensuring a safe, calm, comfortable, and well-designed pedestrian experience in the underground can also serve to encourage modal shift to subway use as well as enforce

[5] Lejk [5].

Fig. 2 Rondo Daszyńskiego, *Source* Metro Warszawskie

the decision of habitual subway commuters to continue using the metro for their commuting needs.

2.1 Warsaw Metro: A Sustainable Urban Mobility Solution

The Warsaw metro is conceived and developed as an integrated public transport system, to promote comfortable and seamless movement across the metropolitan while improving the inhabitants' quality of life. The existing as well as upcoming sections of the Warsaw metro are consistent with the demands of a Sustainable Urban Mobility Plan, which seeks to offer inhabitants a well-planned system with convenient and appropriate routes for their destination (see Fig. 2).[6]

Expansion of the metro system ensures that an increased number of inhabitants (many of whom currently have access to the city's network of buses, trams, and/or urban rapid rail system) will also be able to include metro travel as one of their options for commuting in the city.

A dynamic metropolitan mobility system offers a diversity of options, both for a flexible and agile system as well as to accommodate different preferences and needs.

Warsaw's transport system is based on the Mobility as a Service (MaaS) approach. MaaS integrates various transport infrastructures, complementing existing modes of

[6] Juncà [6].

transport while considering traffic research and the most frequently undertaken trips by passengers. This approach provides passengers with a comprehensive service that combines multiple carriers' offers, navigation systems, and payment technologies into a single, seamless experience.

In Warsaw, these functions are met by the Warsaw City Card system and a range of mobility options including light rail, bus, urban rapid rail, and metro. Passengers have the freedom and flexibility to choose and switch between different modes of transport.

2.2 Exemplary Practices of Walkability: An Overview of Warsaw Metro

The Warsaw Metro boasts numerous design and functional features that can be considered international good practices. Together, these practices contribute to the overall subsurface environment and significantly enhance its walkability (see Fig. 3).

The equipment used throughout the Warsaw metro system follows standardized designs, carefully crafted to meet the observed needs of passengers. Implementations of features like elevators, stairs, and toilets adhere to legal requirements and insights gathered from residents' expectations as well as everyday experiences. Additionally, there is a commitment to continuous improvement, with the Public Transport Authority consistently retrofitting old metro stations. For instance, a recent example includes the addition of an elevator at the Pole Mokotowskie station to enhance accessibility (see Fig. 4).

Some of the key elements observed across Warsaw metro system that enable pedestrian movement across the underground network.

2.2.1 Seating Areas

Ample, practical, and well-designed seating areas are present throughout the city's metro system. Seating is critical in particular for the needs of passengers who might significantly require resting points during their journey but also for the general population (see Fig. 5).

2.2.2 Public Toilets

Public toilets are a sorely neglected amenity in cities across the world and not a common feature in many subway systems.

Access to clean, safe, and free public toilets at every station is a commendable feature of the Warsaw metro system.

1. Every station in the Warsaw metro system has male and female public toilets.

Fig. 3 Elevator to Rondo ONZ, *Source* Metro Warszawskie

2. 2. All public toilets have a staff member or an employee of the contracting company overseeing the management of the toilets particularly the safety, cleanliness, and maintenance during operational hours.
3. All public toilets are free and accessible.

Prioritizing this amenity, Warsaw Metro enhances the walkability of the indoor metro environment as access to clean, safe public toilets is vital for public health and well-being (see Fig. 6).

2.2.3 Accessibility Features

The M1 line of the Warsaw metro was initially designed and implemented with different technical possibilities compared to today. However, accessibility has always been a top priority, leading to the modernization of all older stations to improve accessibility features.

Fig. 4 Tactile tiles, escalator stairs, and path to elevator at Bródno station, *Source* Metro Warszawskie

Fig. 5 seating at Zacisze station, *Source* Metro Warszawskie

Fig. 6 Bemowo station, *Source* Metro Warszawskie

From its inception, the M2 line was designed with a focus on enhancing accessibility for people with disabilities. Stations feature numerous elevators and escalators connecting ground levels with mezzanines and platforms. Tactile maps in Braille are available at stations, aiding orientation for blind individuals. Additionally, designated paths for blind and visually impaired people, marked with white cane guidance, lead from station entrances to various amenities such as elevators, maps, ticket machines, toilets, gates, and platforms. Stair railings also provide Braille information about their destinations. Furthermore, new entry gates without turnstiles have been introduced, offering improved accessibility for individuals using wheelchairs [7].

1. All passenger areas are adapted for differently abled people, the elderly, and people with strollers;
2. The surfaces of the metro stations are smooth, non-slip, and have no thresholds;
3. Elevators are installed across stations to increase the accessibility of stations;
4. Doors at most entrances to stations and metro trains open automatically without constituting a barrier;
5. In the first and last carriage of the train there is a dedicated place for people using wheelchairs, people with small children, and bicycles. By locating them, a passenger can seek help from the driver;
6. New trains are equipped with retractable ramps and buttons to connect with the driver are described in Braille's alphabet;
7. Metro staff undergo periodic training to better provide assistance and cooperation to differently-abled people and procedures for providing assistance and rules of conduct have been developed as well (see Fig. 7).

Fig. 7 Elevator to subway platform, *Source* Metro Warszawskie

2.2.4 Navigation Markings

At each station, there are fluorescent markings navigating passengers toward entrances, exits, elevators, and toilets. Fluorescent pictograms have been introduced as well. Typhlographic plans are placed at the mezzanine and platform levels. Typhlographic plans are special boards that enable blind and visually impaired people to read text or graphics by touch. Moreover, information about the direction and place are indicated in Braille on the handrails of fixed stairs (see Fig. 8).

2.2.5 Safety at Stations and Inside Metro Cars

Ensuring the safety of inhabitants and maintaining a secure environment free from harassment, criminal activity, and other unsafe behaviors is of utmost importance in

Fig. 8 Tactile tiles at Bródno station, *Source* Metro Warszawskie

a subway system. Without a sense of safety, people may hesitate to use underground spaces or systems, regardless of their design quality.[7]

As part of the national police training program, every subway station in Warsaw has two police officers on duty at all times. This active police presence and regular patrolling serve to enforce the law and provide immediate assistance when needed. It also signals to commuters that stations are actively monitored and maintained to ensure a safe environment for all.

2.2.6 Communication System

As a managed indoor environment, elements of communication are vital for the functioning and governance of the space. Employees of the metro system who directly engage with passengers undergo specific training to learn the principles of communication with passengers and provide assistance to people with disabilities.

In terms of audio–video system:

1. There are alarm points at each station, enabling direct audio/video connection with the dispatcher and the central control room;
2. DSO—sound warning system has been introduced at stations and in metro cars;
3. Constant monitoring is carried out at stations and in some metro cars;

[7] Besner [7].

4. At stations with side platforms, the train's entry into the station is preceded by an audio message saying: Attention, a train is coming towards (for example to Kabaty or Młociny);
5. The displays of the Passenger Information System present messages about the waiting time for the next and next train, the type of the approaching train and the next train, text information about the train's arrival at the station for the hearing impaired;
6. The display screens and panels have colors contrasted for the visually impaired.

For telephonic communication, telecommunication equipment is installed at a low height on the gate line to accommodate the needs of people in wheelchairs. There are yellow emergency telephones at each station to contact the station manager. On M1, to facilitate contact with the station staff, additional telephones have been installed at the elevators, adjusted in height to accommodate the capabilities of people in wheelchairs (see Fig. 9).

Fig. 9 Emergency telephone, *Source* Metro Warszawskie

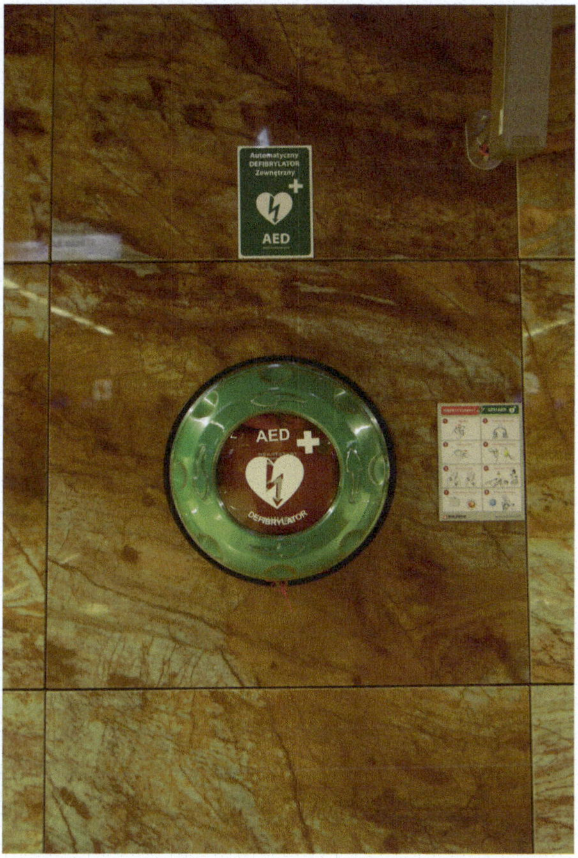

Fig. 10 Defibrillator, *Source* Metro Warszawskie

2.2.7 Medical Support

Defibrillators are available on the platforms of all stations. As with other features standardized through the metro system in Warsaw, this displays an absolute prioritization of safety for travelers and staff and readiness and access to equipment in case of emergencies (see Figs. 10 and 11).

2.3 Warsaw Metro Expansion

The city of Warsaw is rapidly expanding, with its post-industrial areas experiencing rapid residential development and the areas of Kamionek, Gocławek, and Gocław in Praga Południe experiencing particularly intense change. Due to these changes, a decision was made to build another metro line (see Fig. 12).

Fig. 11 Bródno station, *Source* Metro Warszawskie

As of early 2024, the first section of the M3 line is at the design stage. It will be over 8 km long, with 6 stations, and connect the Gocław station with Stadion Narodowy M2 station. The M3 line will be extended under the Vistula River toward Siekierki, and further—through Sielce—to the intersection with the M1 at the Racławicka station. It will run through Wyględów until the Żwirki i Wigury PKP stop, where it will intersect with the M4 (see Fig. 13).

Ultimately, M3 will be a 15-station peripheral line, running in an arc around the center and transporting passengers to other existing lines, the M1 and M2, and to the planned M4, M5.

The master plan indicates that three new lines will be built in Warsaw by 2050. The idea of the 15-min city will in fact be supported by this development, as more than half of the capital's inhabitants are expected to live within walking distance of a metro station, with 17 out of 18 districts having access to the metro.

This development will contribute to the living conditions and quality of life of residents in offering an efficient and fast mode of mobility to navigate the city. The development also enhances and improves the overall walkability of the city.

For further exploration and use of underground spaces for enhanced walkability in Warsaw, it would be wise to invest in:

1. Incorporating elements of light, views, and fresh air to facilitate smoother transitions between underground and surface spaces. This can include strategically placed windows, skylights, and ventilation systems to create a seamless experience for commuters.

Fig. 12 Warsaw metro expansion. *Image source* Metro Warszawskie

2. Providing sensory pleasure and engagement for people using the space. This can be achieved through generous ceiling heights, ample natural sunlight, scenic views of the surroundings, water features, art installations, background music, indoor plants, and pleasant scents, all contributing to a more enjoyable and immersive environment.
3. Promoting subway usage through various media channels and platforms. Highlight the existing connections between stations and surrounding buildings, especially malls and museums, to encourage exploration and increase public awareness of the convenience and accessibility of the subway system. This can involve advertising campaigns, social media outreach, and collaboration with local businesses and cultural institutions to showcase the benefits of utilizing the subway for commuting and leisure activities.

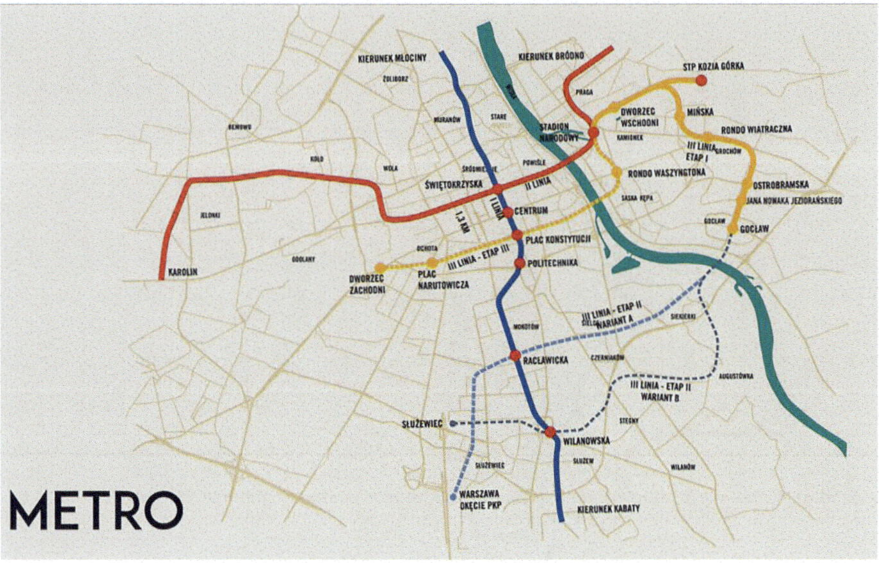

Fig. 13 Master plan for Warsaw Metro 2050. *Image source* Metro Warszawskie

The expansion of the metro system is viewed by Warsaw Metro as a developmental need as well as an opportunity for change. In line with the development of the metro is the parallel transformation of areas and subsequent attracting of new residents. The optimization of space afforded by the metro development lends to the quality of development of the broader area. An interest in investing in the urban space as culturally attractive is also present, with, for example, the placing of graphics by such recognizable artists as Fangor or Młodoczeniec on the walls of underground stops.

Naturally, development concerns in rapidly expanding areas provide a host of other challenging issues for the provision of sustainable mobility options equitably across neighborhoods and districts. We highlight the example of the Warsaw metro system here as possessing the model-worthy elements for a modern, inclusive, and accessible underground.

The authors have no conflicts of interest to declare that are relevant to the content of this chapter.

References

1. UN-HABITAT and ACUUS. https://www.acuus.org/un-habitat/
2. Broere W (2016) Urban underground space: solving the problems of today's cities. TunnLing Undergr Space Technol 55:245–248. ISSN 0886-7798. https://doi.org/10.1016/j.tust.2015.11.012
3. Besner J (Mar 2024) Interview
4. Zacharias J (2023) Underground pedestrian systems, where and how?

5. Lejk J (Mar 2024) Warsaw metro. Interview
6. Juncà J (Mar 2024) Interview
7. Besner J (2017) Cities think underground—underground space (also) for people. Procedia Eng 209:49–55. https://doi.org/10.1016/j.proeng.2017.11.129

Heritage Building Information Modeling (HBIM) and the Digitization of Underground Archaeological Spaces in Amman

Zaid M. Al-Zrigat, Nariman Sami Abdelghani, and Mohammed Raed Ahmed

Abstract This chapter explores the integration of Heritage Building Information Modeling (HBIM) and digital tools in the sustainable utilization of underground archaeological spaces, focusing on downtown Amman, Jordan. The study examines an archaeological site uncovered beneath the city's surface, emphasizing its potential inclusion in urban planning and development strategies. Employing qualitative methods such as thick description analysis alongside advanced digital tools like CerTus HBIM, Revit, Twinmotion, and Photoshop, the research proposes innovative solutions for revitalizing historic urban centers. These solutions include designing staircases and pathways for improved accessibility, relocating critical site elements to enhance preservation, and finalizing the water drainage canal project. The recommendations, developed in collaboration with the Greater Amman Municipality and experts from the Jordanian Department of Antiquities, aim to protect cultural heritage while advancing sustainable urban development in densely populated and rapidly evolving areas such as downtown Amman.

Keywords HBIM · Digitization · Sustainable development · Archaeological · Heritage

---

Z. M. Al-Zrigat (✉) · M. R. Ahmed
Faculty of Engineering, Department of Architecture, Al Al-Bayt University, Mafraq, Jordan
e-mail: zadez6040@yahoo.com

N. S. Abdelghani
Department of Architecture Engineering, The Hashemite University—HU, Zarqa, Jordan

© The Author(s), under exclusive license to Springer Nature Singapore Pte Ltd. 2025
M. Agrawal (ed.), *Underground Spaces for Climate Resilience and Sustainability*, Advances in 21st Century Human Settlements,
https://doi.org/10.1007/978-981-96-0547-7_4

1 Introduction

Cities worldwide face the challenge of balancing population growth, economic development, environmental preservation, and societal well-being. Traditionally, urban planning focuses on land use allocation, including constructing new structures, transportation networks, and recreational areas [1]. However, there is a growing recognition of the need to expand this perspective to include underground spaces due to their potential contributions to sustainable urban development [2].

This chapter addresses several key questions: Why should underground archaeological spaces be considered in urban land use planning? What are the components and resources of these environments? How have some cities integrated archaeological underground space into urban land use planning processes? How can 3D modeling and technology accurately depict underground archaeological spaces in downtown Amman and assist in developing preservation strategies for these sites?

The discovery of archaeological spaces during the construction of a water catchment system in downtown Amman provides an opportunity to reimagine the city's urban landscape, balancing modern development needs with historical preservation. The archaeological site discovered underground includes the remains of a Roman bath, swimming pools, statues, pottery artefacts, and other valuable monuments. Drawing insights from successful models in European cities, the chapter aims to shed light on the potential benefits and challenges of integrating underground spaces into urban planning by emphasizing the importance of utilizing technological innovation, particularly 3D modeling. As shown in Fig. 1.

The chapter focuses on the prototype called "Urban Information Modeling." The model aims to collect all relevant data and knowledge related to the underground archaeological space in the Centre of Amman and update it by proposing alternatives for preservation and reuse processes. This facilitates understanding of this environment and its interaction with the surface, ensuring the rational and effective use of its resources.

2 The Relationship Between the City and Its Archaeological Underground Space

Archaeological sites hold invaluable insights into our collective history, offering glimpses of past civilizations and cultural heritage waiting to be unearthed. As cities expand and evolve, archaeological sites within urban areas face a direct threat of destruction or damage from a variety of emerging environmental and human hazards, making any attempts at their conservation extremely challenging [3].

In some cases, archaeological sites are discovered beneath urban areas following the city's expansion. The utilization of these underground spaces offer opportunities for development, and innovative solutions for conservation, interpretation, and showcasing archaeological sites in a manner that respects their significance, and

Fig. 1 The archaeological site was discovered under the old downtown area. *Source* (authors, 2024)

provides controlled environments while achieving sustainable urban development [4]. The level of integration between the city and its Archeological underground space depends on the Archeological underground surface level, urban planning priorities, financial resources, technological advancements, location within the urban area, and the needs of the community, as shown in Fig. 2.

3 Incorporating the Digital Twin Concept into Archeological Underground Space Documentation

The utilization of digital twins has emerged as a crucial method for documenting and preserving archaeological underground spaces [5]. Unlike traditional statistical sampling methods, this technology offers intricate 3D models of underground structures, fueled by real-time data obtained from sensor systems. This enables advanced visualization, behavioral pattern monitoring, and analysis of archaeological sites, all while reducing the risk of damage to these valuable areas [6].

Archeological Underground Space Documentation and preservation is a comprehensive approach that extends beyond physical conservation to include interpretation, planning, design, construction, and management within the urban contexts [7].

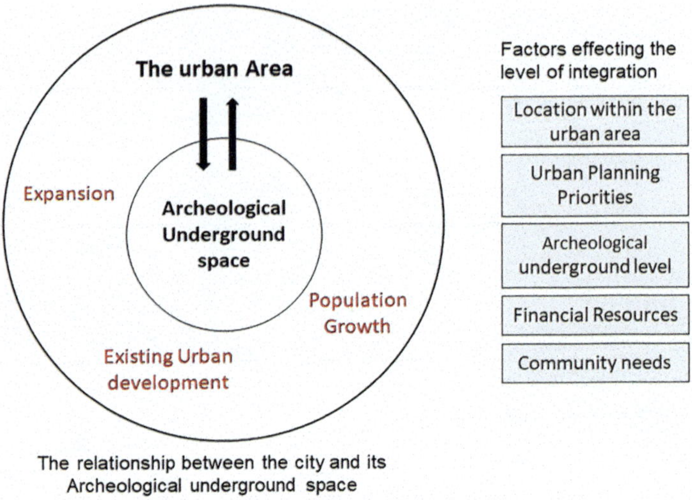

Fig. 2 The relationship between the city and its Archeological underground space. *Source* (authors, 2024)

Through this process, users can provide feedback on various aspects of underground spaces, including layout, accessibility, circulation, and aesthetics.

Furthermore, it facilitates the evaluation of design alternatives and scenarios, allowing planners to assess the potential impacts of different alternatives under various conditions. By incorporating user input into the design process, planners can identify opportunities for underground developments for optimization and enhancement [8].

This approach not only facilitates the visualization and documentation of archaeological underground developments but also enables the exploration of different design alternatives for preserving and integrating these spaces within urban areas before physical construction begins. This process helps identify potential challenges, enhances understanding, and fosters collaboration among various stakeholders, leading to more informed decision-making [9], as shown in Fig. 3.

4 Archeological Underground Space Management for Sustainable Urban Development

Archaeological urban underground space can be used as an opportunity to preserve cultural heritage, but it can also have repercussions on the social, economic, and environmental dimensions of city systems. Archaeological underground spaces may cause displacement in surrounding neighborhoods, leading to social exclusion and loss of community cohesion. In addition, it may affect the preservation and identity of these archaeological sites [1].

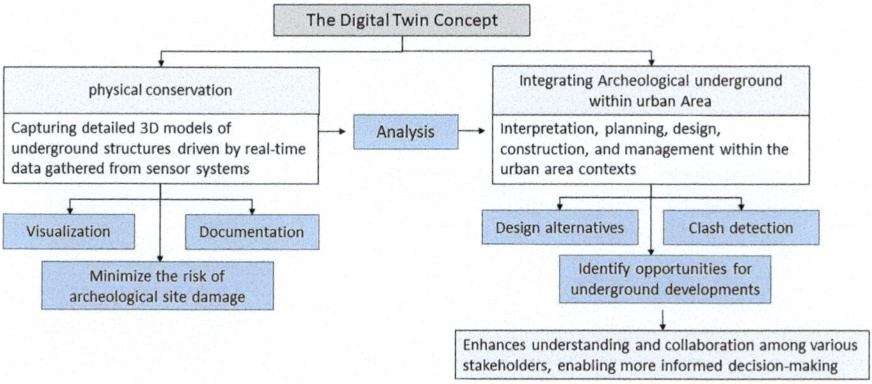

Fig. 3 Incorporating the digital twin concept into archaeological underground space documentation. *Source* (authors, 2024)

Addressing the social challenges requires the implementation of equitable development strategies and a comprehensive approach to preserve the cultural heritage and historical identity, while minimizing disturbance to such sites [10]. In addition, archaeological underground spaces offer opportunities for cultural education and heritage interpretation. Achieving socially sustainable urban development also requires community engagement and participation to ensure that it reflects the needs and priorities of diverse stakeholders [7].

The integration of archeological underground spaces into urban areas may have economic challenges associated with the high cost of preservation, constraints to property development and land use changes [11]. At the same time, addressing the economic challenges can help further tourism [7]. The presence of these underground spaces can also enhance revenue for local businesses and communities, and enhance the value of surrounding properties [12].

The excavation, construction, and maintenance of underground archaeological spaces could have an environmental impact on natural resources, habitat disturbance, biodiversity loss, and waste generation. On the other hand, appropriate integration of the underground archaeological spaces can contribute to habitat restoration and conservation, reduction of air and noise pollution, and reduced energy use and waste generation, as shown in Fig. 4.

4.1 Good Practices of Integrating Underground Archaeological Sites into Urban Fabric

Cities like Rome and Athens have successfully incorporated underground archaeological sites into their urban fabric, allowing visitors to experience historical artefacts while enriching the cultural and historical landscape of the city. By studying these

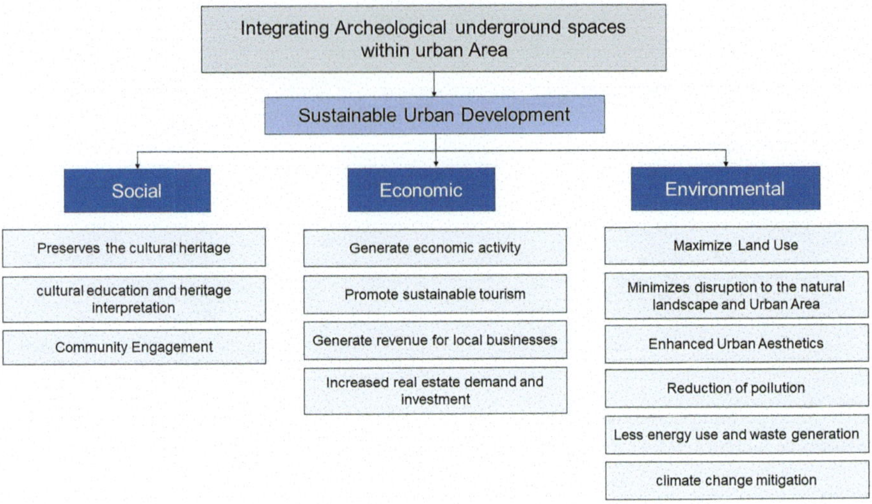

Fig. 4 Integrated approach of sustainable urban development for the Archeological underground spaces. *Source* (authors, 2024)

examples and adapting best practices to the context of downtown Amman, researchers can learn valuable lessons on effectively integrating and sustaining underground archaeological spaces for the benefit of present and future generations.

4.1.1 Piazza Venezia Museum-Station

The project, led by the Metro C consortium, is the first underground project for Rome. It is a highly complex project and a significant engineering challenge not only due to its size and location in the central hub of Rome but also because it is entirely carried out through archaeological excavation [13].

The project will provide a sustainable and efficient connection between the center and the suburbs, with anticipated positive effects on traffic levels and air quality. The work will also include the future development of two museum stations, presenting a unique opportunity to safeguard scattered archaeological sites and remains that are expected to emerge from the excavations and may pose challenges to the construction process. Through recovery work and collaboration with various stakeholders, new models were created for the display of cultural heritage, to preserve the archaeological sites, while taking visitors through a journey into the past [14].

4.1.2 Ancient Streets Beneath the Acropolis Museum in Athens

The Acropolis Museum in Athens Greece contains exhibits in galleries and an open display of underground archaeological site, which was first uncovered in the late

1990s. In 2019, the excavation site was opened to the public. The underground archaeological site contains streets, homes, latrines and baths, workshops, and tombs that date back to the fourth millennium BC. A group of archaeologists, architects, conservators, and engineers came up with a plan to incorporate the underground archaeological site into the museum complex, using two approaches. The first design approach focuses on raised metal walkways that cross the ancient site, giving the impression that it is "floating" above the site. In the second approach, they covered some parts with protective glass, to provide visitors with a unique window into the ancient urban life [15].

5 Methodology

This research focuses on the examination, analysis, and revitalization of the archaeological area that was discovered underground in the old downtown of Amman. The preservation process aims to document the discovered part of the city and develop scenarios for sustainably modernizing and accommodating the needs of local communities. To carry out this preservation effort, a qualitative approach was used to better understand and interpret the archaeological area, and ensure alignment of preservation efforts with historical context and community values [16]. Three key steps of this study include:

1. A comprehensive literature review provides a theoretical foundation and a broad understanding of underground archaeological areas, design processes, and their relationship to heritage preservation. It also explores the principles of sustainable urban planning, offering insights into how these elements can be effectively integrated to preserve cultural heritage while promoting sustainable urban development.
2. Architectural evaluation via on-site surveys, measured drawings, and documentation of spatial characteristics, and architectural components.
3. Modern documentation and simulation tools such as the CerTus HSBIM, Revit, Twin Motion, and Photoshop enabled the digital reconstruction and analysis of the underground archaeological site.

The documentation phase utilized historical archives within the country, government departments, and libraries, including the National Library, which houses political texts and documents. Collaborative efforts with municipal authorities, particularly the Greater Amman Municipality, heritage preservation agencies, and academic institutions were also essential.

Initially, the CerTus HSBIM software was utilized, a cutting-edge program capable of capturing two million points per second. This device provides extensive simulation capabilities and integrates Building Information Modeling (BIM) techniques with urban planning. The software is used to enhance the design and analysis of urban developments and create scenarios to achieve optimal interaction with infrastructure. Operating by generating spatial points represented as a point cloud with coordinates (x, y, z), this method facilitates seamless export to Revit software for creating elevation, plan, and 3D models [17].

For the archaeological underground area, advanced BIM land modeling was adopted, including direct import of land profile files from Google Maps with satellite images. Data collection was facilitated using Cyclone FIELD software on an iPad connected to CerTus HSBIM, allowing data extraction with coordinates for various objects [18]. The original software was rented from the Freedom Survey Consulting Office in Amman, supported and guided by specialized engineers in the office.

The researchers used a plugin called BIM Voyager to incorporate different scenarios derived from the overall view of the underground archaeological site. Each scenario was linked to a specific timeline to illustrate various situations in time and place to stakeholders. Subsequently, the original archaeological site file, along with the proposed alternatives, was exported in IFC format and imported into Autodesk Revit for remodeling the site with precise dimensions, as shown in Fig. 5.

Autodesk Revit was chosen for its ability to professionally construct archaeological site elements, create detailed drawings and maps, and export them in various formats such as DXF, DWG, CAD, BMP, and PICT. The researchers' methodology,

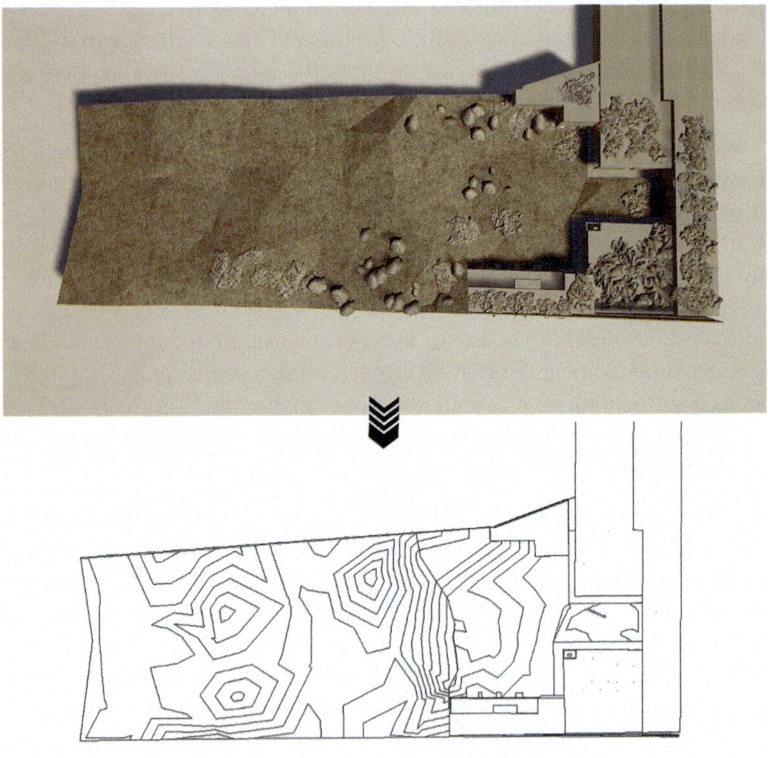

Fig. 5 Export the archaeological site in IFC format and share the model with us via the BIM cloud platform. *Source* (authors, 2024)

which includes the use of CerTus HSBIM software for documentation and scenario design, is shown in Fig. 6.

Utilizing the REVIT architectural program, a virtual site mirroring its real-world counterpart was constructed. It highlighted the site's dimensions, materials, and other architectural elements. This enabled the generation of elevations, projections, and accurately proportioned doors and windows. Subsequently, the final output was exported to AutoCAD for additional refinement of line weights and dimensions. Following the completion of documentation processes, Twin Motion and Photoshop helped merge images and text, and enhance old historical images for 3D manipulation.

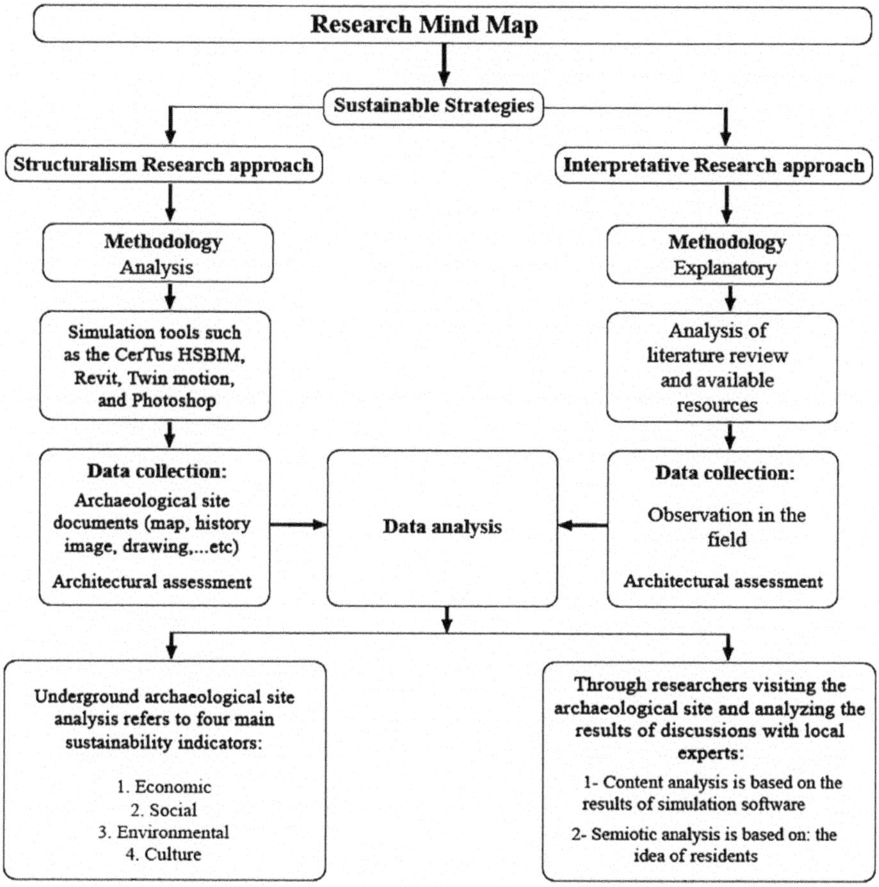

Fig. 6 Research mind map. *Source* (authors, 2024)

6 Key Findings

Accurately documenting architectural elements using digital tools like CerTus HSBIM and Revit allows researchers to enhance the cultural significance of the underground archaeological site. This underscores the importance of its preservation and subsequent revitalization to serve the needs of the local community.

The architectural style of the discovered sections reflects Roman influence, akin to structures found in neighboring regions such as Karak and Salt. This influence is evident in the design of the bathroom, featuring semi-circular arches with varying dimensions and longitudinal arches standing at different heights. Additionally, sedimentary stone, readily available in the region, is utilized as the primary building material.

These structures exhibit a symmetrical design and straightforward composition, reflecting the architectural standards of their era without overt symbolic significance. By meticulously documenting these architectural features and their impacts, researchers gain valuable insights into strategies for reactivating and repurposing these spaces while preserving their historical significance and addressing contemporary needs. As shown in Fig. 7.

Engaging local communities and architectural practitioners is essential in shaping the quality of urban practices in the old downtown area. Collaborating with governments and integrating their visions and preferences into the underground reuse project can transform the Roman Baths area into not only a cultural landmark but also a vibrant hub for community engagement and heritage appreciation.

The plan to repurpose the underground archaeological site encompasses a series of suggestions and proposals tailored to the financial capabilities of the Jordanian local authorities. These initiatives aim to introduce various activities while preserving

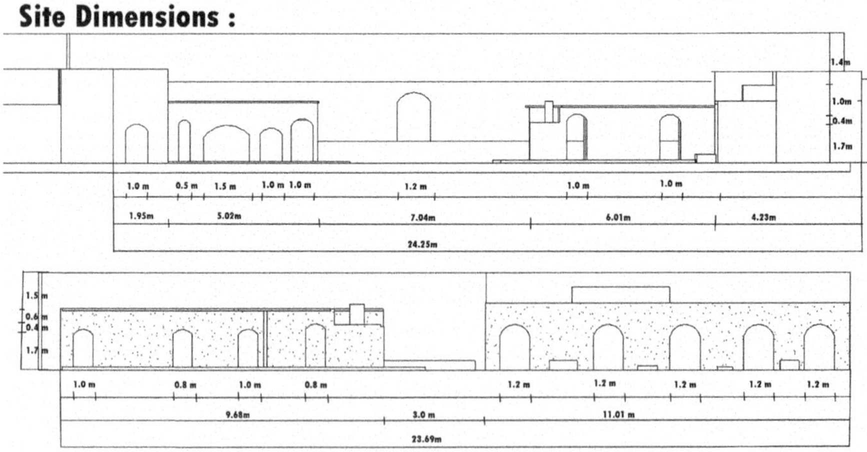

Fig. 7 The documentation part of the archaeological site using CerTus HSBIM and Revit programs. *Source* (authors, 2024)

Fig. 8 3D modeling of the basic case of the archaeological site. *Source* (authors, 2024)

the architectural and heritage essence of the site. Researchers utilized an integrated suite of programs, including CerTus HSBIM, Revit, and Twin Motion, to meticulously document the original case of the underground archaeological site upon its discovery, as shown in Figs. 8, 9, and 10. Subsequently, they developed proposals and modifications at both architectural and urban levels to rejuvenate and conserve the site. These proposals include:

1. Adding staircases and pathways for movement inside and around the archaeological site.
2. Transforming the archaeological site into a museum by covering it with reinforced glass, creating a walkable passage for visitors.
3. Linking the archaeological site with the Roman Theater, emphasizing Amman's significance as an ancient cultural center. The Theater, being the largest in Jordan, holds a pivotal place in the downtown area's cultural heritage.
4. Transferring a portion of the archaeological site to the National Museum and organizing events and activities in its place.
5. Incorporating staircases and movement pathways both inside and around the archaeological site would enhance accessibility and navigation for visitors, facilitating their exploration of the historical treasures buried beneath downtown Amman.

Transforming the archaeological site into a museum by covering it with reinforced glass creates a walkable passage for people, enhancing accessibility and preserving the historical significance of the site. Opening the site to the public as a museum can significantly boost tourism in the area, attracting visitors eager to explore ancient civilizations. This influx of tourists brings positive economic impacts, generating revenue for local businesses and creating job opportunities. Moreover, the museum

Fig. 9 The first proposal involves modifying the original case by adding corridors both around and inside the archaeological site. *Source* (authors, 2024)

serves as an educational resource, offering valuable insights into the daily lives, customs, and technologies of the ancient Romans. It also provides research opportunities for scholars interested in further studying the site, thus contributing to the advancement of archaeology and history. The use of reinforced glass allows natural light to illuminate the space, reducing the need for artificial lighting and promoting energy efficiency. Overall, transforming the archaeological site into a museum creates a communal space for learning and interaction, fostering a sense of pride and unity within the local community (Fig. 11).

Linking the underground archaeological site with the Roman Theater highlights Amman's importance as an ancient cultural hub and can attract tourists, bolstering the local economy. As the largest and oldest theater in Jordan, the Roman Theater

Fig. 10 The first proposal involves modifying the original case by adding corridors under the archaeological site. *Source* (authors, 2024)

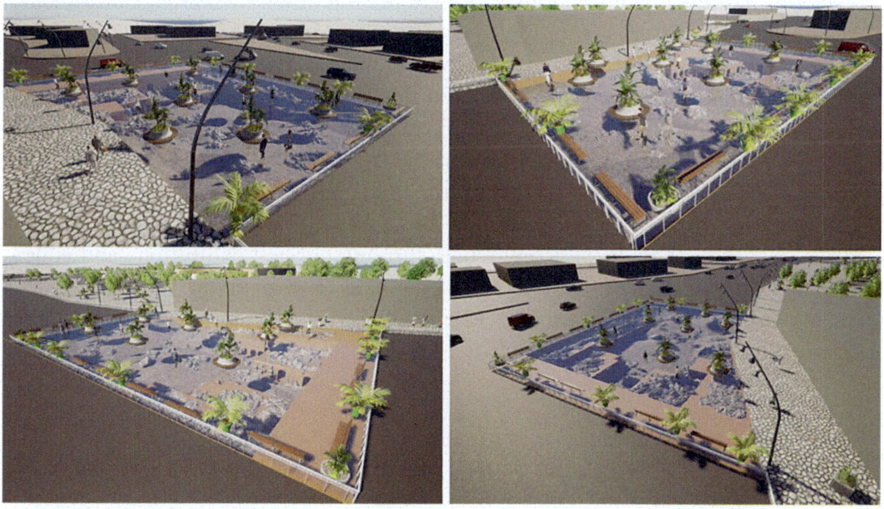

Fig. 11 The second proposal involves modifying the original case by covering the archaeological site with reinforced glass, mirroring international case studies. *Source* (authors, 2024)

holds significant historical value. By creating detailed 3D models of the underground archaeological spaces and integrating them with the Roman theater, researchers can

Fig. 12 The third proposal involves modifying the original case by connecting the archaeological site with the Roman theater in the downtown area. *Source* (authors, 2024)

conduct interdisciplinary studies, uncovering new insights into ancient urban planning, architecture, and daily life. Additionally, these digital resources serve as valuable educational tools, allowing students and scholars worldwide to remotely engage with Amman's heritage, as indicated in Fig. 12.

Transferring a portion of the underground Roman archaeological site, particularly the Roman bath, to the National Museum and organizing events and activities around it serve multiple crucial purposes. Firstly, it addresses the recreational needs of residents and fills a gap in public spaces in the old downtown area. Additionally, moving delicate structures from the underground site to a controlled environment like the National Museum ensures their long-term preservation and protection. The controlled climate and security measures in the museum help prevent deterioration caused by exposure to the elements or human activity, safeguarding these valuable historical structures for future generations.

Furthermore, by relocating part of the archaeological site to the National Museum, it becomes more accessible to the public. Visitors can learn about the history and significance of the Roman bath through curated exhibitions and educational programs. This accessibility promotes greater awareness and appreciation of Amman's cultural heritage, fostering a sense of connection and pride among locals and tourists alike (Fig. 13).

Fig. 13 The fourth proposal involves amending the original case by transferring part of the archaeological site to the National Museum and organizing events and activities instead. *Source* (authors, 2024)

7 Conclusion

The integration of archaeological heritage into Amman's urban revitalization and economic diversification efforts is crucial for shaping the city's identity. Urban masterplans should seamlessly incorporate underground archaeological spaces into downtown Amman to foster sustainable economic growth and community engagement. This integration can catalyze job creation and transform deteriorating urban areas into vibrant tourist destinations.

Through our study, we aimed to analyze the rationale behind utilizing archaeological urban underground spaces, identify components and resources, and explore their integration into urban planning. By focusing on the Roman archaeological site in downtown Amman and drawing insights from case studies in European cities, we proposed a stock model for urban information gathering, sharing, and updating to ensure proper utilization.

Currently, the integrated management of Greater Amman Municipality's resources lacks a comprehensive approach, leading to the wastage of Urban Underground Space (UUS) resources. Addressing competition for UUS resources requires long-term land use planning and an integrated assessment of infrastructure. Sustainability in the master plan for the Old Downtown area can be achieved through three-dimensional planning and an integrated assessment of infrastructure.

Our research pioneers innovative approaches to digital documentation and the creation of various simulation scenarios using tools like CerTus HSBIM and Revit. These efforts set a precedent for similar endeavors globally, showcasing the potential of digital tools in preserving and integrating archaeological heritage into urban planning.

Competing Interests: The authors have no conflicts of interest to declare that are relevant to the content of this chapter.

References

1. Zerhouny M, Fadil A, Hakdaoui M (2018) Underground space utilization in the urban land-use planning of Casablanca (Morocco). Land 7(4):143
2. Bobylev N (2009) Mainstreaming sustainable development into a city's Master plan: a case of urban underground space use. Land Use Policy 26(4):1128–1137
3. Ping Z et al (2009) On utilization of underground space to protect historical relics model. Tunn Undergr Space Technol 24(3):245–249
4. Cui J, Broere W, Lin D (2021) Underground space utilisation for urban renewal. Tunn Undergr Space Technol 108:103726
5. Cureton P, Dunn N (2021) Digital twins of cities and evasive futures. Shaping smart for better cities. Elsevier, pp 267–282
6. Ivanov S et al (2020) Digital twin of city: concept overview. In: 2020 global smart industry conference (GloSIC). IEEE
7. Erincik S (2021) Integration of archaeological sites into planning process: the case of Iskele (Urla). Izmir Institute of Technology (Turkey)
8. Chandran R, Chon H (2021) Redefining heritage and cultural preservation through design: a framework for experience design. In: Congress of the international association of societies of design research. Springer
9. Ghodsvali M (2018) 3D modelling of underground space for urban planning and management-providing basic planning insight. University of Twente
10. Çalışkan M, Hamamcioglu-Turan M (2021) Integration of archaeological heritage in Metropolitan city centres with urban life: Agora of Smyrna. Conserv Manag Archaeol Sites 23(1–2):6–30
11. Mubaideen S, Al Kurdi N (2017) Heritage conservation and urban development: a supporting management model for the effective incorporation of archaeological sites in the planning process. J Cult Herit 28:117–128
12. Montazerolhodjah M, Pourjafar M, Taghvaee A (2015) Urban underground development an overview of historical underground cities in Iran 25(1):53–60
13. di Grazia I (2020) Metro C, surveys completed: there is an approval for the PIazza Venezia station. LEGGO
14. Matteis S (27 Oct 2023) What the new Venezia station of Rome's Metro C will be like. FanPage
15. Howitt-Marshall D (2022) Athens underground: the ancient streets below the acropolis museum. https://www.greece-is.com/athens-underground-the-ancient-streets-below-the-acropolis-museum/
16. Creswell J (2008) Planning, conducting, and evaluating quantitative and qualitative research, 4th ed. Universitas of Nebrasaka, Lincoln: Pearson

17. Martins ALG, Lima LSSD, Santos ET (2021) O uso do BIM no canteiro de obras e a sua influência na segurança do trabalho. Anais SBTIC
18. Mayer P et al (2021) Analysis of the current state of automation of hazard detection processes in BIM in Slovakia. Appl Sci 11(17):8130

Climate and Carbon: The Subterranean Solution

Towards Sustainable Underground Development Through Synergies

Fabienne Sierro and Katrin Pakizer

Abstract The importance of underground resources in addressing societal challenges such as climate change, population growth, and urbanisation is increasingly evident, as they play a crucial role in advancing sustainability goals. However, current underground development is often uncoordinated and reactive, following ad-hoc, "first-come, first-served", or last-resort principles. This fragmented approach results in unsustainable management practices that overlook opportunities for synergies between underground uses. This chapter outlines synergistic relationships between various underground resource uses by drawing on real-world examples. We have identified three core principles that enable subsurface synergies: multifunctionality, circularity, and repurposing. By showcasing examples from around the globe, we illustrate how these principles can be integrated into legal planning frameworks, such as master plans, to promote sustainable underground development and ensure long-term functioning. To facilitate and implement subsurface synergies effectively, a holistic, interdisciplinary, and integrated approach to underground planning is essential. Researchers, policymakers, legislators, and practitioners must go beyond mere conflict resolution arising from competing underground uses to leverage synergies for long-term sustainable development above and below ground.

Keywords Sustainable underground development · Synergies · Sustainability · Underground resources · Subsurface planning · Subsurface management

---

F. Sierro (✉) · K. Pakizer
Institute of Innovation and Entrepreneurship, ZHAW, Theaterstrasse 17, 8401 Winterthur, Switzerland
e-mail: sieo@zhaw.ch

K. Pakizer
e-mail: paki@zhaw.ch

F. Sierro
Department of Environmental Systems Science, Transdisciplinarity Lab, ETH Zurich, Universitätstrasse 16, 8092 Zurich, Switzerland

K. Pakizer
Department of Environmental Social Sciences, Swiss Federal Institute of Aquatic Science and Technology (Eawag), Ueberlandstrasse 133, 8600 Dübendorf, Switzerland

© The Author(s), under exclusive license to Springer Nature Singapore Pte Ltd. 2025
M. Agrawal (ed.), *Underground Spaces for Climate Resilience and Sustainability*, Advances in 21st Century Human Settlements,
https://doi.org/10.1007/978-981-96-0547-7_5

1 Introduction

The underground contains physical space, groundwater, geothermal energy, and geomaterials, which are essential resources for countries to become resilient and sustainable [1, 2]. Various societal challenges require solutions which can, in part, be provided by these underground resources. For instance, the underground is increasingly being considered in climate change mitigation and energy transition strategies, due to its supply of natural resources such as heat, gas, space for storage, and water [2, 3]. The underground also provides space for infrastructures (e.g., underground railway systems) and subsurface construction projects. There is a pressing demand for such space as population growth and urbanisation continue and require solutions in the domain of mobility and construction to unburden the surface from excessive traffic, congestion, and density stress [4, 5]. The underground space can also be used as a safe space for infrastructures and populations facing extreme weather conditions (e.g., droughts, windstorms, tsunamis, floods, etc.) due to climate change [6–8]. Moreover, it has been estimated that value derived from underground resources could help achieve eleven of the 16 UN Sustainable Development Goals [9].

As societies and different sectors (e.g., construction, transportation, energy) increasingly recognise the potential of the underground, different subsurface uses compete for limited suitable geological formations. Allocating the rights of use to a particular underground development project has long-term consequences for other use options, creating path dependencies that potentially impede future developments [10–12]. Until now, the underground has been characterised as a chaotic, unregulated space where ad-hoc, "first-come, first-served", or last-resort principles apply [10, 13–15]. In other words, underground development has been primarily concerned with relocating infrastructures below ground, usually without a long-term vision for the underground space or its resources [16]. Current governance arrangements reinforce this issue by regulating underground uses separately within policy-making silos. The resulting lack of coordination and integration increasingly complicates underground constructions and constrains access conditions for future development projects.

As the demand for underground resources continues to grow, it becomes imperative to adopt a more holistic management approach to ensure sustainable underground development. Sustainability in itself should not merely be a goal advanced by leveraging underground resources; it should also be the guiding principle for their use and management.

Sustainable underground management aims to ensure the equitable distribution of these limited resources across generations, preventing the depletion of individual resources through excessive exploitation. Additionally, it seeks to coordinate the use and protection of subsurface resources to achieve long-term social, environmental, and economic goals.

By identifying and prioritising sustainable practices, we can better manage underground resources to address contemporary challenges while preserving their availability and utility for future generations. This integrated approach ensures that underground development contributes to overall sustainability objectives, with particular relevance in the context of urban environments.

Most research has shed light on the role and potential of underground space to contribute to sustainability goals by comparing subsurface uses with above-surface uses, discussing their advantages and disadvantages [8]. There has also been some research on the synergies between underground and above-ground uses (i.e., Ref. [17]). However, limited attention has been given to the interactions between underground uses and their associated resources with respect to underground sustainability. Moreover, most research has focused on conflicts between underground uses [4], while synergistic relationships between these uses have only been marginally addressed. Research has yet to provide comprehensive findings on how sustainable underground development can be achieved, and particularly to what extent synergies between underground uses may contribute to sustainability.

Debrock et al. [3] formulate a first set of comprehensive design recommendations for an urban sustainable underground space based on an integrative literature review. They argue that four aspects need to be considered for sustainability in urban underground spaces: livability, connectivity, temporality, and interdependency. Livability refers to the subsurface as a living space for people and the environmental services that support human well-being and comfort. Connectivity describes the need to transition from solitary ad-hoc underground management to an interconnected and coordinated network approach. Temporality refers to limited underground resources that are non-renewable and require future-proof developments. Finally, interdependency highlights existing and potential synergistic links between underground uses and use flows.

This chapter focuses on the last key aspect of sustainable underground design identified by Debrock et al. [3], i.e., the synergistic interdependencies. Drawing on examples of urban and rural contexts across the world and a review of research, we outline how synergies between underground uses may contribute to sustainable underground development. Additionally, we put forward three main principles on which underground synergies are based. Finally, we discuss key factors for promoting and applying subsurface synergies.

2 Synergies for Sustainable Underground Development

Synergies between underground uses are beneficial interactions among subsurface resources, technologies, and/or infrastructures. In other words, a synergy arises when one type of use enables or promotes another type of use. Opportunities to use such synergies are often overlooked due to the lack of strategic and holistic underground development. However, synergies between underground uses can significantly contribute to sustainable underground management and, thereby, to sustainable cities and regions [9]. This section outlines the benefits of using subsurface synergies for sustainable underground development.

A key benefit is optimising resource use while minimising environmental impact. Underground developments tend to impact each other more significantly than above-ground developments due to factors such as the geological properties of the subsoil. Most conflicts between underground uses arise from density stress, where different uses require access to or occupy the same geological layers. Synergies can mitigate this density stress by integrating various uses into multifunctional infrastructures, thus avoiding infrastructure duplication, resource wastage, and decreasing environmental disturbance. Such multifunctional infrastructures often require less surface area, freeing up surface space and reducing the visual impact.

Subsurface synergies further enhance cost efficiency of building and operating infrastructure. Underground construction costs are often two to four times higher than the costs of constructing similar-scale projects above-ground [18]. However, in most cases, underground constructions prove to be economically viable when all long-term costs (i.e., both initial and operation costs) are considered [8]. As land costs decrease with depth, overall project costs can be reduced, especially in cities and regions with high land prices [19]. At the same time, real estate with underground public infrastructure may increase in value by providing a better transportation system and environmental quality [8]. Underground spaces are naturally protected from external influences and have a stable temperature, resulting in minimal maintenance expenses [13]. At the project level, coordinated underground construction projects can profit from economies of scale via shared services, bulk purchases, optimised distribution and transportation of resources, and lower per-unit costs for equipment and materials.

Harnessing synergies between underground resources necessitates extensive coordination and interdisciplinary planning. By fostering interdisciplinary collaboration, subsurface synergies can fuel innovation, which in turn can lead to a competitive advantage for products and services. Thereby, different actors work towards shared value creation. Additionally, community engagement plays a vital role in this process, ensuring that underground development aligns with local residents' and stakeholders' needs and aspirations. Particularly in the context of sustainability, interdisciplinary collaboration is crucial when tackling complex real-world challenges and identifying shared values to move from business-as-usual policies to more sustainable configurations.

Finally, utilising synergies between underground uses may alleviate the pressure on above-ground space and provide secure and protected spaces for critical infrastructure and resources. For example, by integrating telecommunication lines in underground tunnel systems, the above-ground can be alleviated from traffic and visual nuisances. Moreover, such infrastructures are more resilient to climate change when situated in the subsurface. For example, Hanamura [6] and Vähäao [7] emphasise that earthquakes and tsunamis in Japan have caused little structural damage to underground transportation infrastructures, while such events greatly damage aboveground infrastructures. As climate change will increasingly lead to such extreme weather conditions, the synergistic use of underground resources will potentially become more relevant.

3 Principles of Subsurface Synergies: Multifunctionality, Circularity and Repurposing

Exploring synergies between different underground uses offers pathways to sustainable underground management, impacting social, economic, and environmental dimensions. Drawing from our research in the Swiss context [20] and practical examples, we identify synergies emerging primarily along three principles: multifunctionality, circular economy principles, and repurposing initiatives. Table 1 provides a summary of the three principles with an illustrative example for each. While not exhaustive, the following subchapters serve as a foundational resource for future research and aid practitioners in recognising the synergy potential of their underground projects.

3.1 Multifunctionality

The principle of multifunctionality refers to uses that can serve multiple purposes simultaneously or interchangeably [21]. In the context of the underground, this principle is often encountered in subsurface transport infrastructures. Tunnels, in

Table 1 Overview of the three core principles of underground synergies with an illustrative example

Type of synergy	Description	Illustrative example
Multifunctionality	Subsurface uses that serve multiple purposes simultaneously or interchangeably	Tunnel infrastructure for mobility with integrated pipelines for various sectors, such as electricity, water, district heating and cooling, telecommunications (Example: Helsinki, Finland)
Circularity	Subsurface uses promoting the reuse, recycling, and repair of subsurface materials and products	Subsurface tunnel construction with direct recycling and reusing of excavated geomaterials (Example: Lötschbergtunnel, Switzerland)
Repurposing	Former subsurface uses transformed to serve a new purpose once their original function is fulfilled or no longer required	Abandoned mines repurposed into water reservoirs for energy storage and geothermal energy production (Example: Heerland, the Netherlands)

particular, are increasingly being conceived as multi-utility tunnels, integrating other underground uses and transporting goods and/or people.

Following an integrated planning and management approach, the city of Helsinki exemplifies the integration of pipelines and cables of various sectors in tunnel infrastructure. Lying 30–80 m below ground level, Helsinki has a 60 km long network of multi-purpose tunnels combining roadways for vehicular access with cables and pipelines for electricity, water, telecommunications, district heating, and cooling [5, 22]. In line with this idea, pedestrian utility tunnels function similarly by combining pipelines and cables for various uses while providing pedestrians with a route, for example, to avoid heavy traffic above-ground [1, 23].

The SMART tunnel (Stormwater Management and Road Tunnel) in Malaysia stands as the world's longest multi-purpose tunnel. It serves two purposes: alleviating traffic congestion and managing excess water during extreme weather events above-ground. Depending on the volume of water, the tunnel operates in two stages: initially, a bypass tunnel redirects water, ensuring the flow of traffic, while subsequently, both transportation and bypass tunnels facilitate floodwater drainage [24]. Paris is another example of a multifunctional tunnel system combining utility uses with sewer systems.

Tunnels are not solely dedicated to passenger traffic; they also play an important role in freight transport. A notable example is Cargo Sous Terrain in Switzerland, a pioneering large-scale underground freight transportation system. With ambitions to traverse the nation, it seeks to interconnect different cities via an extensive underground tunnel network. While primarily focusing on freight transportation, Cargo Sous Terrain has explored the possibility of transporting CO_2 for storage alongside goods and materials. A feasibility study revealed that incorporating CO_2 transportation presents a promising avenue for the multifunctional use of the freight transportation system [25].

Catering to both regular and emergency needs, such as those arising from wars or extreme weather events, multi-purpose underground facilities exemplify adaptability in usage. Itäkeskus in Helsinki, for example, is an underground swimming pool primarily serving as a leisure and sports venue, accommodating up to 1000 customers at a time. However, in times of crisis, this subterranean space can transform into an emergency shelter, providing refuge for approximately 3,800 people. Such versatile underground structures demonstrate the importance of flexible design and utilisation, ensuring functionality in various scenarios to enhance community resilience [7]. Some other notable examples include Gangnam International Transit Center in Seoul, South Korea, combining transport infrastructure with commercial and leisure zones; and Gjøvik Olympic Cavern Hall in Norway for accommodating the Winter Olympic games while providing space for defence.

3.2 Circularity

Circular economy approaches, centred on sustainable resource management, waste reduction, and resource efficiency, have become increasingly relevant for achieving sustainability objectives. These promote reusing, recycling, and repairing materials and products, fostering a closed-loop economic system. In the underground realm, various resources and uses offer potential synergistic connections based on circular principles, often involving the recovery and further utilisation of waste materials.

One established synergy, grounded in recycling and reusing principles, is evident in the relationship between underground construction and the excavation of geomaterials. Subsurface projects often require substantial quantities of excavation, which can be recycled and reused for construction purposes. Countries like Austria, Switzerland, and France recycle much of the excavated materials, minimising environmental impacts by processing them on-site or at nearby dedicated facilities. For instance, the Lötschberg Base tunnel in Switzerland, completed in 2007, entirely self-supplied its construction site with recycled aggregates from excavation activities [26].

Additionally, circular economy principles facilitate synergies between underground infrastructures (e.g., tunnels, subsurface parking) and geothermal energy production. Energy geostructures exemplify this synergy, harnessing the residual heat from underground constructions to generate energy. Usually, such geostructures can be integrated into conventional structural elements (such as walls, tunnel linings, piles, etc.) [1] or added to walls [27]. This approach maximises resource efficiency and contributes to sustainable energy production, further highlighting the potential of circular economy practices in the underground domain. Another example of efficient use of "waste" heat are subsurface data farms, where the heat produced by operating data servers is distributed to the district heating network [17].

Geothermal energy and water exploitation also bear the potential for circular use. Geothermal energy and potable/non-potable water can be gained from groundwater. After harnessing the geothermal energy (for heat or electricity production) from groundwater, the remaining water can be desalinated and used for irrigation in the agricultural sector or be treated to provide fresh drinking water. Examples of this procedure can be found in the Polish Lowlands (central part of Poland) and the southern part of the country, where low-enthalpy geothermal waters are being treated to serve as drinking water after providing district heating [28, 29]. However, treating such geothermal water for the purpose of water supply requires additional energy. To ensure sustainability principles, this additional energy should be provided by renewables. Another circular synergy emerges between geothermal energy and salt, as evident in the Reykjanes geothermal area in Iceland. Here, groundwater naturally mixes with seawater, resulting in highly saline water. Through a vacuum evaporation technique, salt brines and process steam can be extracted from geothermal wells [30]. This process enables commercial production of common salt for household use. Additionally, geothermal water can serve lithium extraction in certain cases, offering a more environmentally friendly alternative to traditional lithium mineral exploitation [31]. These synergies underscore the potential of geothermal energy to facilitate sustainable resource utilisation and circular economy practices.

Finally, the subsurface also provides space for storing energy, goods, and global warming gases. Recent advances in carbon capture, storage and utilisation technologies enable the mineralisation of CO_2 into basalts and peridotites [32]. This mineralisation process occurs in the shallow underground and may potentially "produce" geomaterials, which, in turn, could be used for construction purposes.

3.3 Repurposing

Underground infrastructure can become obsolete due to various factors, such as economic viability, environmental concerns, or shifting societal needs. In such cases, sites may be abandoned or dismantled, paving the way for repurposing them for new functions. Repurposing involves transforming sites and infrastructures originally designed for specific underground uses to serve a new purpose once their original function is fulfilled or no longer required.

Quarries, for instance, are excavated into the shallow underground to extract raw materials such as sand, gravel, limestone, or granite. Once the extraction process is complete, quarries often resemble empty pits and can be repurposed into landfill sites. An example is the transformation of the Mellikoner quarry in Switzerland from a limestone extraction site into a landfill for uncontaminated excavation material, with the aim of re-naturalisation. It illustrates how obsolete underground infrastructure can be repurposed to foster sustainable land use and environmental restoration. Another example stems from the Swiss canton Aargau. The concession process for a quarry site required a strategy for its repurposing once its function was fulfilled. Consequently, the project developers collaborated with the national supermarket chain Coop to develop a logistics centre, built into the former quarry.

Abandoned mines offer opportunities for repurposing, often transforming into subsurface water reservoirs for various purposes. These sites can naturally fill with water or be artificially filled, serving as reservoirs for geothermal energy production or energy storage. For example, in Heerland, the Netherlands, abandoned coal mines now provide heating and cooling to the town [17]. Water-filled mines can also function as underground pumped hydroelectric energy storage, generating electricity on demand. Additionally, these groundwater reservoirs can be jointly utilised for water supply and geothermal energy production, as demonstrated in successful cases in China [33].

Moreover, abandoned mines can serve as storage spaces for energy production through compressed air rather than water. Deep geothermal boreholes also hold potential for repurposing as deep geological repositories for the safe disposal of contaminated waste, such as nuclear waste. While research on this multifunctional repurposing of deep geothermal energy and deep geological storage is still limited, it lays the groundwork for potential future synergistic uses.

4 Challenges in Leveraging Synergies for Sustainable Underground Development

The implementation of subsurface synergies for sustainable underground development poses multifaceted challenges. Identifying, promoting, and implementing such synergies requires cross-sectoral collaboration, holistic governance, and innovative approaches. Existing regulatory frameworks often focus on avoiding conflicts rather than incentivising synergies, resulting in a lack of evaluation of synergistic interdependencies during authorisation processes for subsurface projects. Sectoral planning approaches further exacerbate this issue, hindering coordination and exchange between administrative bodies responsible for environmental, social, and economic aspects. Moreover, the three-dimensional nature of subsurface planning is often overlooked until the implementation phase of spatial planning, complicating integrated management efforts.

Many projects struggle to use their waste or by-products simply because their permits were only granted for single resource use and purpose. An example is the district heating system in Riehen, Switzerland, which produces a significant quantity of salt from the geothermal waters at 1545 m below ground. This salt could be reprocessed to supply table salt following the circular economy-based principle. However, their concession permit does not allow for the economic valorisation of their processual "waste" and would require an additional authorisation process.

A lack of knowledge about the underground further complicates efforts to identify and implement synergies. Countries without an oil and gas industry, in particular, need to first explore their underground and its geological properties. Comprehensive data collection and centralised data-sharing platforms are essential for unlocking synergistic opportunities.

Additionally, while holistic management requires cross-sectoral collaboration and multidisciplinary approaches, complex legal procedures and economic pressures often hinder it. Time is valuable, and collaboration is often perceived as an additional cost and time constraint. However, research shows that early collaboration and participation results in longer-lasting, more legitimate, and accepted projects, which are economically viable in the long run as they reduce legal disputes [34].

Master plans have the potential to facilitate synergies by raising awareness of interdependencies, collaborative potential, and encouraging multifunctional, circular project designs. A few cities, such as Helsinki and Singapore, have developed comprehensive master plans for subsurface management, addressing legal challenges and coordinating between various interests. These successful cases demonstrate the benefits of participative, interdisciplinary planning processes. By fostering stakeholder engagement and cross-sectoral collaboration, these cities have overcome challenges and unlocked synergistic interdependencies, paving the way for sustainable underground development.

5 Discussion and Conclusion

The underground harbours crucial yet underutilised resources for advancing sustainability. Recognising its potential across sectors and policy domains, a departure from ad-hoc planning and "first-come first served" management is imperative. This chapter emphasises that sustainable underground development, contributing to broader sustainability objectives, necessitates integrated, holistic, and interdisciplinary approaches.

Synergistic interdependencies between underground uses offer substantial potential to support sustainable development, foster sustainable cities and regions, and facilitate cross-sectoral collaboration. These synergies optimise resource utilisation, minimise environmental impact, and promote cost efficiency. Achieving them demands collaboration and innovation across sectors, disciplines, and policy silos, prioritising integrated, holistic, and long-term planning aligned with societal needs and future resource preservation.

As elucidated in this chapter, subsurface synergies can be fostered through multifunctionality, circular economy principles, and repurposing initiatives. Examples worldwide span from integrated infrastructure to resource reprocessing and repurposing of abandoned facilities. However, research on subsurface synergies remains limited and necessitates expansion for a comprehensive understanding.

Unlocking such synergies entails addressing challenges like knowledge gaps, limited data-sharing, interdisciplinary collaboration barriers, and fragmented governance and legal frameworks. Cases like Helsinki and Singapore exemplify the efficacy of early subsurface integration into spatial planning frameworks, facilitated by legally binding and holistic master plans. These plans should explicitly incorporate synergies and advocate for permitting processes conducive to multifunctional, circular, and repurposing-oriented project designs.

Future research must identify pathways to facilitate and implement subsurface synergies, offering policymakers and practitioners actionable recommendations. By doing so, we can unlock the underground's full potential for sustainable development while ensuring its preservation for future generations.

References

1. Li H, Li X, Soh CK (2016) An integrated strategy for sustainable development of the urban underground: from strategic, economic and societal aspects. Tunn Undergr Space Technol 55:67–82. https://doi.org/10.1016/j.tust.2015.12.011
2. Volchko Y, Norrman J, Ericsson LO, Nilsson KL, Markstedt A, Öberg M, Mossmark F, Bobylev N, Tengborg P (2020) Subsurface planning: towards a common understanding of the subsurface as a multifunctional resource. Land Use Policy 90:104316. https://doi.org/10.1016/j.landusepol.2019.104316
3. Debrock S, Van Acker M, Admiraal H (2023) Design recommendations for sustainable urban underground spaces. Tunn Undergr Space Technol 140:105332. https://doi.org/10.1016/j.tust.2023.105332

4. Li X, Li C, Parriaux A, Wu W, Li H, Sun L, Liu C (2016) Multiple resources and their sustainable development in Urban Underground Space. Tunn Undergr Space Technol 55:59–66. https://doi.org/10.1016/j.tust.2016.02.003
5. Delmastro C, Lavagno E, Schranz L (2016) Underground urbanism: master plans and sectorial plans. Tunn Undergr Space Technol 55:103–111. https://doi.org/10.1016/j.tust.2016.01.001
6. Hanamura T (2011) Open session in world tunnel congress 2011. Helsinki
7. Vähäaho I (2016) An introduction to the development for urban underground space in Helsinki. Tunn Undergr Space Technol 55:324–328. https://doi.org/10.1016/j.tust.2015.10.001
8. Zargarian R, Hunt D, Rogers C (2013) The role of underground space in a sustainable UK urban environment. In: Zhou C, Sterling (eds) Advances in underground space development. The Society for Rock Mechanics & Engineering Geology (Singapore), pp 249–250. ISBN: 978-981-07-3756-6
9. Peng F-L, Qiao Y-K, Sabri S, Atazadeh B, Rajabifard A (2021) A collaborative approach for urban underground space development toward sustainable development goals: critical dimensions and future directions. Front Struct Civ Eng 15:20–45. https://doi.org/10.1007/s11709-021-0716-x
10. Bartel S, Janssen G (2016) Underground spatial planning—perspectives and current research in Germany. Tunn Undergr Space Technol 55:112–117. https://doi.org/10.1016/j.tust.2015.11.023
11. Bauer S, Beyer C, Dethlefsen F, Dietrich P, Duttmann R, Ebert M, Feeser V, Görke U, Köber R, Kolditz O, Rabbel W, Schanz T, Schäfer D, Würdemann H, Dahmke A (2013) Impacts of the use of the geological subsurface for energy storage: an investigation concept. Environ Earth Sci 70:3935–3943. https://doi.org/10.1007/s12665-013-2883-0
12. Von Der Tann L, Sterling R, Zhou Y, Metje N (2020) Systems approaches to urban underground space planning and management—a review. Undergr Space 5:144–166. https://doi.org/10.1016/j.undsp.2019.03.003
13. Bobylev N (2009) Mainstreaming sustainable development into a city's Master plan: a case of Urban Underground Space use. Land Use Policy 26:1128–1137. https://doi.org/10.1016/j.landusepol.2009.02.003
14. Admiraal H, Cornaro A (2016) Why underground space should be included in urban planning policy—and how this will enhance an urban underground future. Tunn Undergr Space Technol 55:214–220. https://doi.org/10.1016/j.tust.2015.11.013
15. Stones P, Heng TY (2016) Underground space development key planning factors. Procedia Eng 165:343–354. https://doi.org/10.1016/j.proeng.2016.11.709
16. Parriaux A, Tacher L, Joliquin P (2004) The hidden side of cities—towards three-dimensional land planning. Energy Build 36:335–341. https://doi.org/10.1016/j.enbuild.2004.01.026
17. Brown H (2014) Next generation infrastructure: principles for post-industrial public works. Island Press, Washington, Covelo, London
18. Zhao Z, Cao Q (2011) The development of urban underground space from the perspective of urban economy. Procedia Eng 21:767–770. https://doi.org/10.1016/j.proeng.2011.11.2076
19. Hanamura T (1990) Japan's new frontier strategy: underground space development. Tunn Undergr Space Technol 5:13–21. https://doi.org/10.1016/0886-7798(90)90058-R
20. Sierro F, Meyer C, Pakizer K, Abegg A (2024) Untergrundnutzung in der Schweiz: Potenziale, Konflikte und Regulierungsbedarf. Swisstopo, 1–62
21. Finstad J, Dahl Andersen A (2023) Multi-sector technology diffusion in urgent net-zero transitions: niche splintering in carbon capture technology. Technol Forecast Soc Chang 194:122696. https://doi.org/10.1016/j.techfore.2023.122696
22. Chow FC, Paul T, Vähäaho IT, Sellberg B, Lemos LJL (2002) Hidden aspects of urban planning: utilisation of underground space
23. Boivin DJ (1990) Underground space use and planning in the Québec City area. Tunn Undergr Space Technol 5:69–83. https://doi.org/10.1016/0886-7798(90)90062-O
24. Kannapiran RKM (2005) A study and evaluation on SMART project, Malaysia
25. Cargo sous terrain AG (2023) Noch mehr Klimaschutz dank Cargo sous terrain. https://www.cst.ch/wp-content/uploads/2023/07/230705_MM_CST_D.pdf

26. Voit K, Murr R, Cordes T, Zeman O, Bergmeister K (2020) Tunnel spoil recycling for concrete production at the Brenner Base tunnel in Austria. Struct Concr 21:2795–2809. https://doi.org/10.1002/suco.202000434
27. Laloui L, Rotta Loria AF (2020) Analysis and design of energy geostructures: theoretical essentials and practical application. Elsevier-Academic Press, London
28. Tomaszewska B, Akkurt GG, Kaczmarczyk M, Bujakowski W, Keles N, Jarma YA, Baba A, Bryjak M, Kabay N (2021) Utilization of renewable energy sources in desalination of geothermal water for agriculture. Desalination 513:115151. https://doi.org/10.1016/j.desal.2021.115151
29. Tomaszewska B, Pająk L, Bundschuh J, Bujakowski W (2018) Low-enthalpy geothermal energy as a source of energy and integrated freshwater production in inland areas: technological and economic feasibility. Desalination 435:35–44. https://doi.org/10.1016/j.desal.2017.12.032
30. Kristjánsson I (1992) Commercial production of salt from geothermal brine at Reykjanes, Iceland. Geothermics 21:765–771. https://doi.org/10.1016/0375-6505(92)90029-9
31. Murodjon S, Yu X, Li M, Duo J, Deng T (2020) Lithium recovery from brines including seawater, Salt Lake Brine, underground water and geothermal water. In: Vizureanu P (ed) Thermodynamics and energy engineering. IntechOpen (2020)
32. Snæbjörnsdóttir SÓ, Sigfússon B, Marieni C, Goldberg D, Gislason SR, Oelkers EH (2020) Carbon dioxide storage through mineral carbonation. Nat Rev Earth Environ 1:90–102. https://doi.org/10.1038/s43017-019-0011-8
33. Zhang M, Xie Z, He L (2022) Does the scarcity of urban space resources make the quality of underground space planning more sustainable? A case study of 40 urban underground space master plans in China. Front Environ Sci 10:966157. https://doi.org/10.3389/fenvs.2022.966157
34. Gunton TI, Day JC (2003) The theory and practice of collaborative planning in resource and environmental management. Environments 31:5–20

Utilising the Subsurface for Sustainable Green Energy Solutions

Chrysothemis Paraskevopoulou, Alex Connolly, Timothy Kearsey, and Nicholas Shaw

Abstract The rapid migration from rural to urban areas is leading to overpopulation in cities, posing significant challenges for urban planners. By 2050, it is estimated that 70% of the global population will reside in urban areas. Governments are striving to expand cities to accommodate this growth while preserving green spaces and ensuring a high quality of life. This task is complicated by the limited availability of space. Underground space, often perceived solely as a domain for tunnels, utilities, and mining, offers immense potential for urban environments. This misconception neglects the evolution of underground space into more complex and diverse applications. One innovative application of underground space is shallow geothermal energy schemes utilising abandoned mine workings. This approach offers a low-carbon heating solution crucial for the UK's goal of achieving net-zero emissions by 2050. It is particularly relevant in historical industrial cities with extensive coal mining voids. This study examines the geothermal potential of coalfield mine workings in Greater Leeds, providing a methodology for establishing geothermal energy opportunities in similar geological settings. The insights and methods derived from this study are applicable not only in the UK but also internationally, showcasing the transformative potential of underground space in addressing contemporary urban challenges.

Keywords Underground space · Geothermal energy · Minewater heat recovery · Coal mining · Underground space sustainability · Net zero · SDGs

---

C. Paraskevopoulou (✉) · A. Connolly · N. Shaw
School of Earth and Environment, University of Leeds, Leeds, UK
e-mail: c.paraskevopoulou@leeds.ac.uk

T. Kearsey
British Geological Survey, Edinburgh, UK

© The Author(s), under exclusive license to Springer Nature Singapore Pte Ltd. 2025
M. Agrawal (ed.), *Underground Spaces for Climate Resilience and Sustainability*, Advances in 21st Century Human Settlements,
https://doi.org/10.1007/978-981-96-0547-7_6

1 Introduction

Modern cities are increasingly focused on cultivating urban ecosystem—to respond to multifaceted challenges linked to rapid population growth. A major obstacle is the limitation of space above ground, which restricts horizontal city expansion in places like Hong Kong and New York City. Moreover, horizontal expansion of cities isn't sustainable; it can negatively affect citizens' well-being by increasing commute times and greenhouse gas emissions, thus harming the environment. Additionally, spreading out reduces the efficiency of infrastructure and energy use [15].

Underground spaces are an often-overlooked component of urban areas. To address surface space constraints, it is imperative to integrate underground spaces into urban planning and development [14]. By examining urban transformations through the lenses of critical urbanism, geography, and cultural studies, we can understand the complexities of cities and urban regions. The dynamic structuralism approach introduced by Yiftachel [20] suggests that contemporary urbanism is shaped by multiple simultaneous forces, movements, agents, and politics. This approach underscores the importance of recognising the dominant structural forces in shaping cities, particularly those related to economic life, globalisation, and capitalistic space production.

This research aims to answer three key questions: How can urban underground spaces become essential parts of cities and be fully integrated into their development? How can cities incorporate underground spaces into their multi-layered dimensions to achieve organic growth? Can underground spaces contribute to sustainable development by providing an example of Urban Shallow Geothermal—utilising abandoned and flooded underground mines in the greater area of Leeds in the UK, thereby contributing to the resilience and sustainability of smart cities?

2 Underground Spaces and Sustainability

Underground space is often undervalued in urban environments. While commonly associated with tunnels, utilities, and mining activities, it also supports building foundations, agriculture, and mineral extraction, all of which are non-renewable resources [15].

Underground spaces are often perceived as dark, unwelcoming, and even frightening. This misconception overlooks the fact that in densely populated areas, the most frequented public spaces are often underground metro lines and their extensive stations, some reaching depths of up to 100 m. As noted by Paraskevopoulou et al. [14] it's essential to shift public perception to view underground spaces as inviting and friendly environments that promote social interaction and inclusion. Today, underground spaces (Fig. 1) are utilised for various purposes, including geothermal energy, groundwater resources, storage facilities, parking, underground farming, and public recreational facilities like swimming pools and sports centres. However, this expansion has been sporadic and not well-planned, leading to issues of overcrowding and misuse. Recognising these issues early has highlighted the need for comprehensive planning of subsurface use, though this understanding is not yet widespread among urban policymakers. The goal now is to plan underground space effectively and sustainably, incorporating innovative ideas to maximise its potential for future generations.

According to Paraskevopoulou et al. [14], urbanisation processes necessitate a thorough understanding, reevaluation, and redesign of underground spaces. These spaces are becoming increasingly crucial in urban transformation, especially in cities struggling with overpopulation and high density. The concept of the "right to the city," defined as the "right to change ourselves by changing the city," extends beyond aboveground urban areas to include adjacent underground spaces.

According to Paraskevopoulou et al. [14, 15], historical records indicate that ancient civilisations, such as the Greeks, Egyptians, and Romans, faced environmental challenges such as deforestation, salinisation, and soil degradation. However, despite early awareness of environmental concerns, it wasn't until much later that

Fig. 1 Examples of user-friendly and sustainable use of underground space in the UK. (Photos Courtesy: C. Paraskevopoulou)

Fig. 2 Sustainable development goals introduced in the 2030 agenda for sustainable development proposed by the United Nations members in 2015 (U.N., 2019—publicly available)

sustainable development became a topic of discussion in literature, notably introduced by John Mill in 1843. Since then, various authors have contributed their perspectives on sustainable development.

In modern times, the widely quoted definition of sustainable development was articulated by the World Commission on Environment and Development in 1987. The Commission defined sustainable development as "development that meets the needs of the present without compromising the ability of future generations to meet their own needs." Nearly thirty years later, in 2015, all United Nations Member States adopted the 2020 Agenda for Sustainable Development, which introduced 17 Sustainable Development Goals (SDGs), as depicted in Fig. 2. Within these goals, 169 targets were established, aiming to integrate social, economic, and environmental dimensions of sustainable development. These agendas aspire to create "a world free of poverty, hunger, disease, and deprivation," where living environments are safe, resilient, and sustainable, with equitable access to affordable, reliable, and sustainable energy sources [21]. Many of these SDGs are relevant to the development of underground spaces and are considered in the work presented here.

2.1 Alignment with Global Goals

Underground space use can help realise various Sustainable Development Goals (SDGs) [14]. To delve deeper into this topic, Paraskevopoulou et al. [15] created a database of 30 underground projects that was scrutinised and categorised based on

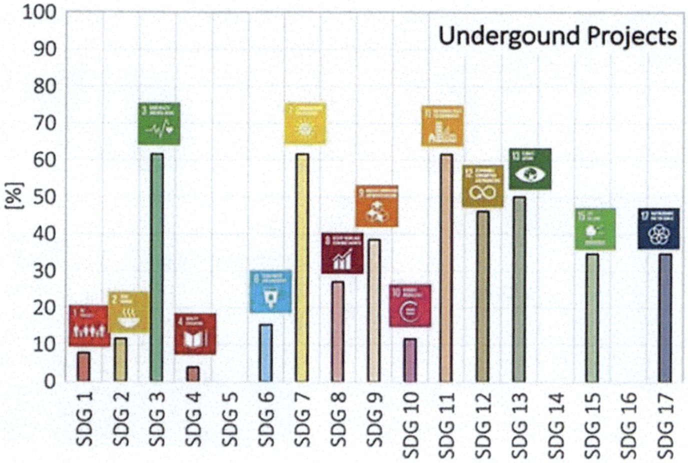

Fig. 3 SDGs and underground projects [15]

their intended purposes into the following groups: transportation and commuting, energy (including geothermal and hydropower), storage (such as waste storage), recreational, housing, and food and beverage (FnB).

The analysis revealed that a majority of the SDGs can be addressed through underground projects. Specifically, as shown in Fig. 3, 66.6% of the examined projects aimed to contribute to SDGs 3, 7, and 11, focusing on promoting good health and well-being, providing affordable and clean energy, and fostering sustainable cities and communities. Additionally, half of the projects (=15) targeted SDG 13, which pertains to climate action.

Figure 4 provides illustrative examples of such underground spaces. Particularly noteworthy is the transition to underground spaces, where even semi-underground housing (depicted in Fig. 4c) can significantly reduce energy demands. Benardos et al. [5] conducted a comparative study on the total energy demand of a typical aboveground house and an earth-sheltered building on Kea Island in Greece (with a climate similar to the Attika region. The findings revealed that the surface building had an overall energy demand 42% higher than that of the underground building.

2.2 *Shallow Geothermal*

Amid climate crisis, decarbonising energy supply is vital. Often overlooked in the energy decarbonisation discussions is space heating. Shallow geothermal energy offers a sustainable, low-carbon solution for heating, contributing significantly to net-zero targets and overall sustainability. Leveraging this renewable resource is essential for creating resilient, eco-friendly urban environments [14].

Fig. 4 Sustainable Underground Projects: **a** Underground boat parking/storage, **b** Underground Green Farming and **c** Underground Housing, Paraskevopoulou et al. [15]

Compared to traditional high enthalpy geothermal energy, which relies on groundwater temperatures exceeding 100 °C for power generation or above 50 °C for direct heating, shallow geothermal energy operates at lower enthalpy levels and can provide usable space heating from groundwater at temperatures as low as 10 °C [8, 13]. A particular form of shallow geothermal energy, known as mine water geothermal, utilises water stored in flooded, abandoned coal mines as its heat source.

Since the Industrial Revolution, increased coal demand led to extensive mining, particularly around industrial hubs in the Midlands, Northern England, South Wales, and Scotland. Many major industrial cities are thus situated above abandoned mines. While mine water geothermal has been successfully piloted in countries like the UK, Netherlands, Spain, Canada, and the USA, large-scale city-wide schemes are rare, limiting our understanding of their full potential.

This research evaluates abandoned mines across the Greater Leeds region to ascertain the feasibility and thermal potential of implementing mine water geothermal schemes. The objectives include assessing whether such schemes are: (a) feasible, (b) beneficial in terms of promoting greener energy and providing social benefits, particularly to disadvantaged households aligned with net-zero goals, and (c) feasible using existing data. Datasets from the Coal Authority, coupled with published estimates of Yorkshire's geothermal gradient, were used for this purpose. The analysis involved determining the spatial distribution, total volume, and temperature of mine water within the workings, and modelling selected areas to assess long-term stability. QGIS, supplemented by a digital terrain model, facilitated the analysis of workings polygons and the determination of their depth below ground level [8, 13].

2.2.1 Heat Generation and Storage

The urgency to address and achieve UK's 2050 net-zero target puts pressure on communities to act fast [10]. One important possibility is to utilise existing infrastructure and man-made geo-assets like abandoned mine working. For this purpose, it is crucial to understand the potential of these existing geo-assets. One of the main challenges is a rapid, and relatively cheap assessment not requiring additional data acquisition [4]. This section highlights the potential challenges and opportunities that need to be considered for evaluating the best step forward in utilising mine workings for thermal heating.

Shallow geothermal energy is defined by the MUSE (Managing Urban Shallow Geothermal Energy) research project, as heat extracted from or injected into the subsurface for storage or exchange, typically at a depth of upto 200 m below ground level [9]. Equation 1 indicates how power output can be determined from this heat extraction [13]:

$$\text{Thermal Power(watts)} = Q * \rho_w * c_w * \Delta T \tag{1}$$

where: Q = pumping rate (m^3/s), ρ_w = density of water (kg m^{-3}), c_w = specific heat capacity of water (J kg^{-1}), ΔT = temperature difference for energy extraction (°C)

Implementing shallow geothermal energy systems can be done via closed or open loop systems, and can be particularly effective in urban settings. These systems can be applied alongside district heating networks or for heating individual buildings of various sizes. However, like most energy generation methods, shallow geothermal schemes benefit from economies of scale.

Key components of a shallow geothermal heating system include the heat pump and heat exchanger. Groundwater extracted from the subsurface serves as a "cold source." It enters the heat pump to evaporate a refrigerant into vapour. The vapourised refrigerant passes through a compressor to raise its temperature, and the resulting superheated vapour transfers heat to a secondary hot water circuit or "hot source" via a heat exchanger. The refrigerant's temperature is then reduced using an expansion valve before cycling back to the beginning of the loop.

To fully assess the heat extraction potential of a source, such as legacy mine workings, it is necessary to understand the volume of water that the mine galleries can accommodate and the temperature difference that can be extracted by the system based on the subsurface temperature capacity.

2.3 *Mine Water Geothermal in Abandoned Coal Workings*

There are three key methods to extract thermal energy from mine water (closed loop, open loop, and standing column), of which open loop is the most commonly used across the UK [1, 2, 8, 12]. Both open and closed systems have variations in execution, mainly in terms of where the heat exchanger is installed and whether or

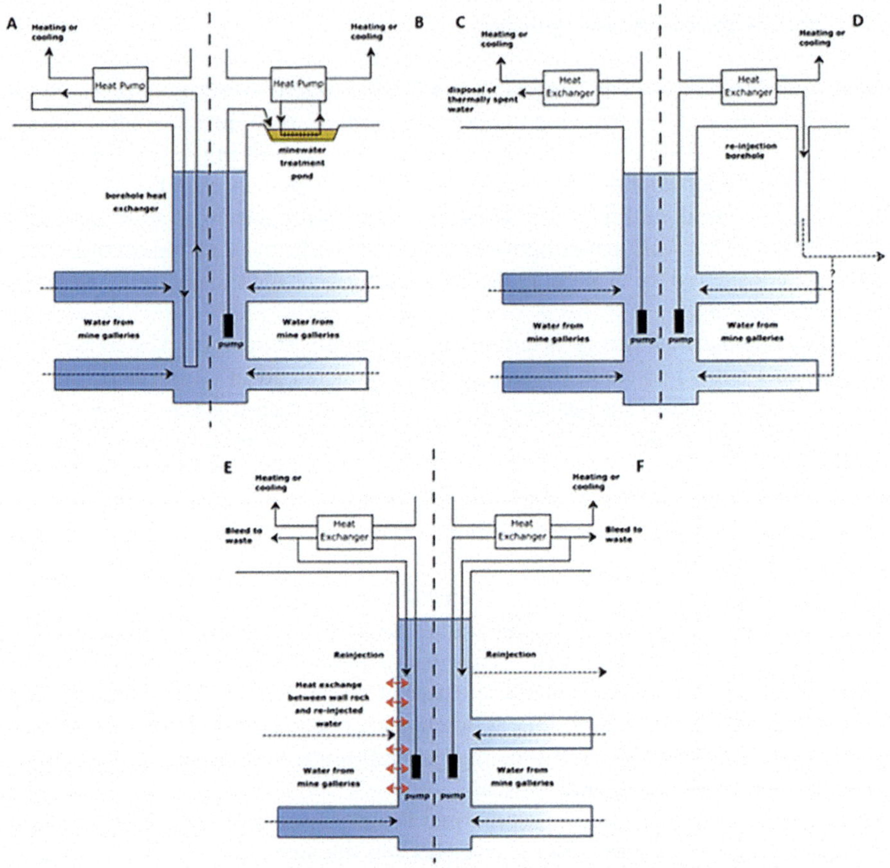

Fig. 5 A and B: Closed loop mine water geothermal system with heat exchanger submerged in either (a) mine shaft, or (b) mine water treatment pond. C and D: Open loop system with either (c) disposal of thermally spent water, or (d) re-injection. E and F: Standing column system with (e) water flowing up or down the shaft and exchanging heat with the wall rock, or (f), flowing back into the mine galleries (modified from Ng et al. [13])

not the thermally spent mine water is re-injected (Fig. 5). Table 1 enlists existing projects and corresponding type of heat extraction system in use.

2.3.1 Heat Flow Within Flooded Mine Workings

The heat extracted from mine water geothermal projects is primarily determined by the regional geothermal gradient beneath the site. Farr et al. [8] created a map of the geothermal gradient beneath Britain's coalfields and found that the average gradient is 24.1 °C per km. However, this can vary across regions from 17.3 °C per km to 34.3 °C per km, with coalfields in the Northeast of England, Cumbria, and

Table 1 Examples of operational mine water geothermal projects [3, 7, 11, 16, 18] Open loop 1 = disposal of pumped water, Open loop 2 = re-injection of thermally spent water

Name	Location	System type	Inflow water temperature (°C)	Delivered temperature (°C)
Caphouse Colliery	Wakefield, Yorkshire	Closed loop and open loop 1	14.5	
Markham No. 3 Shaft	Sheffield, Yorkshire	Standing column	14–15	52–55
Shettleston	Glasgow, Scotland	Open loop 2	10	
Barredo-Figaredo mine workings	Asturias, NW Spain	Open loop 1	>20	46–50
Herleen	The Netherlands	Open loop 2	hot = 28; cold = 16	45
Name	Location	System type	Inflow water temperature (°C)	Delivered temperature (°C)
Springhill	Nova-Scotia, Canada	Open loop 2	18	
Dannenbaum mine	Bochum, W Germany	Seasonal storage	~13	
Dawdon	Newcastle	Open loop 1	19	

Yorkshire showing the highest geothermal gradients of over 30 °C per km. These values represent equilibrium conditions and can be changed locally by mine water pumping. At the depths utilised for shallow mine water geothermal (~100–300 m), expected equilibrium temperatures in the UK range from roughly 12 to 20 °C [12].

3 Area of Interest: The Greater Leeds Case in the UK

The study under investigation focuses on the Greater Leeds Area in the UK outlined in Fig. 6.

3.1 Geology

The geology of northwest Yorkshire, east of the Pennines, is dominated by Carboniferous rocks, divided into three main groups: Carboniferous Limestone, Millstone Grit, and Coal Measures. Leeds itself lies above the northern section of the East Pennine Coalfield and is primarily underlain by the Pennine Lower and Middle Coal Measures Formations, with some outcropping of underlying Millstone Grit Group

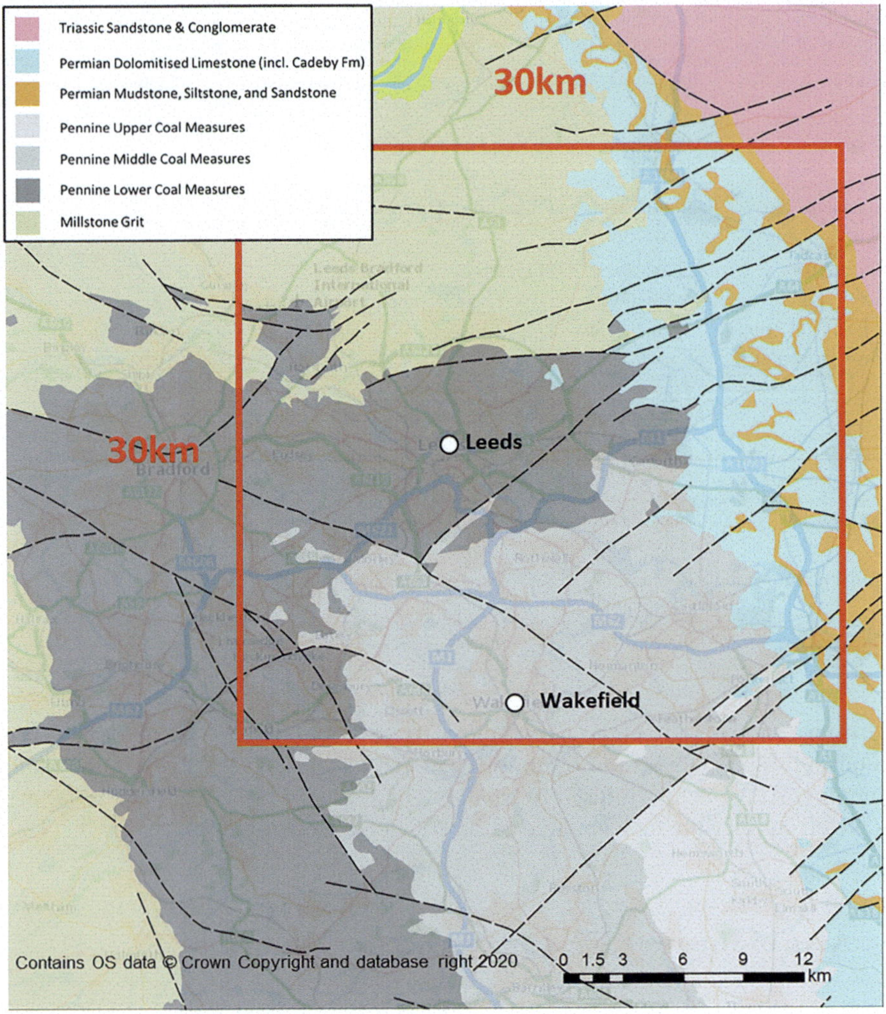

Fig. 6 Broad geology of greater Leeds area (study area) (public data from British Geological Survey, ©BGS, UKRI 2021)

to the north, as well as the overlying Permian strata of the younger Permian aged Cadeby Formation to the east (Fig. 6) [19].

3.2 Coal Seams

The Coal Measures are the most critical strata within the study area. They are made up of interbedded dark grey mudstone/shale, siltstone, and pale grey sandstone, containing marine bands and coal seams, as well as associated clay ironstones [6, 19]. The lower boundary of the Pennine Lower Coal Measures Formation is marked by the Subcrenatum Marine Band, which lies conformably over the Millstone Grit Group, with the upper limit separating it from the Pennine Middle Coal Measures formation demarked by the Vanderbeckei Marine Band. The interbedded sandstones from laterally discontinuous channel bodies within the Lower and Middle Coal Measures. The most notable channel systems within the study area include the Elland Flags within the Pennine Lower Coal Measures Formation and the Thornhill Rock at the base of the Pennine Middle Coal Measures Formation, which represent different rock masses from strong to weak (or weathered) [17].

The Yorkshire coalfield contains over 50 worked seams across the span of the Lower and Middle Coal Measures Formations, although the lack of standardised naming conventions means that record of seam names are not always consistent on a local to regional-scale. Post-WW2, under the aegis of the National Coal Board, names became more consistent, but to this day there are still discrepancies between BGS and Coal Authority seam designations.

3.3 Geothermal Gradient

According to Farr et al. [8], the average geothermal gradient across Yorkshire's coalfields is 32.3 °C per km, which is roughly 8.2 °C more than the average calculated for Britain's coalfields. This value was calculated using measurements of equilibrium geothermal gradient at 100 m intervals between 100 and 1100 m depth. These geothermal gradients were calculated using mine water temperature data from 5 to 10 mine water blocks within Yorkshire, as delineated by the Coal Authority. Figure 7 depicts the variability of average mine water temperature measured within the relevant Yorkshire mine water blocks.

4 Methodology

This work primarily focuses on GIS analysis. The flowchart in Fig. 8 provides an overview of the case study. The GIS three-step analysis aims to interpret data provided by the BGS to assess the shallow geothermal potential of all mine workings across the study area and identify high-potential areas for detailed modelling. This approach (Fig. 8) can be applied to establish the foundational understanding of geothermal energy prospects in analogous geological contexts, and leverage mining legacies beyond the UK.

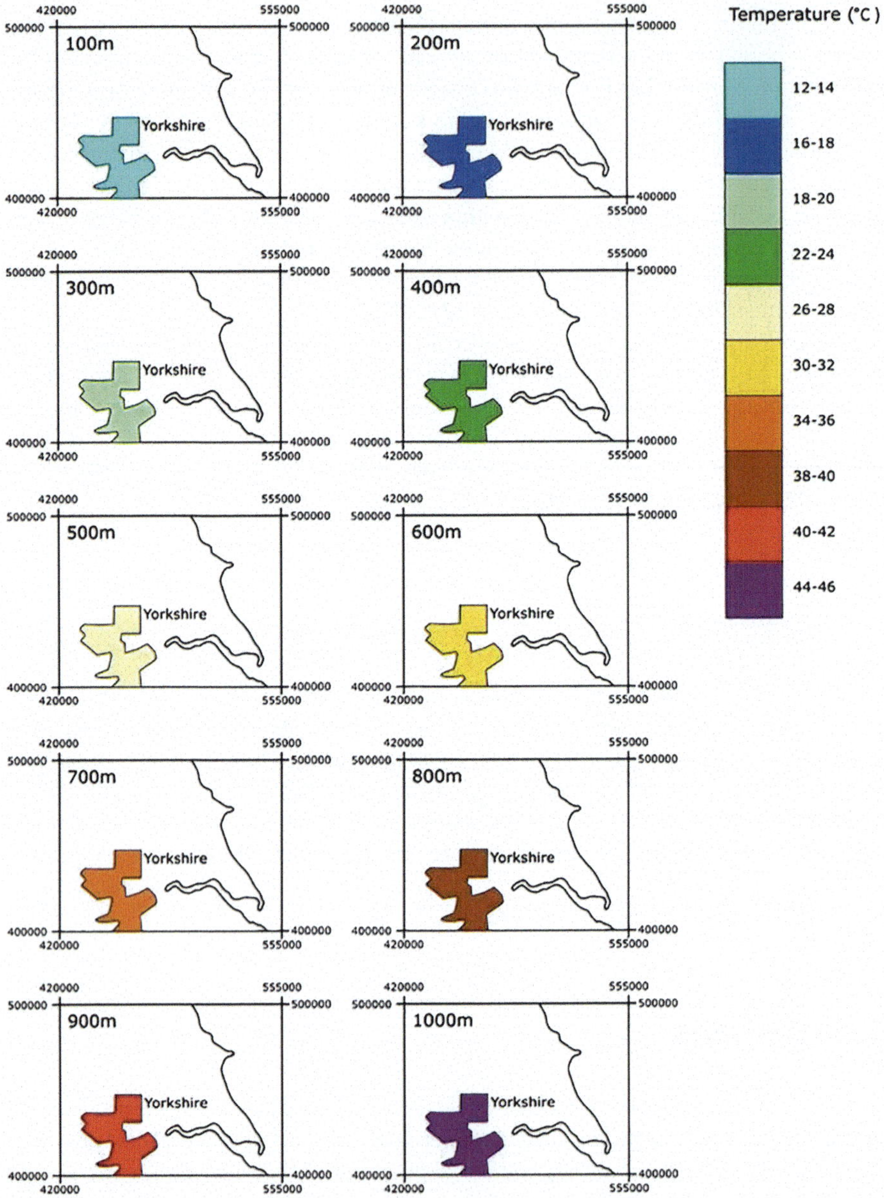

Fig. 7 Average mine water temperature at 100 m depth intervals within the Yorkshire mine water block, determined from a combination of down-borehole measurements and historical data (modified after Farr et al. [8]—public data)

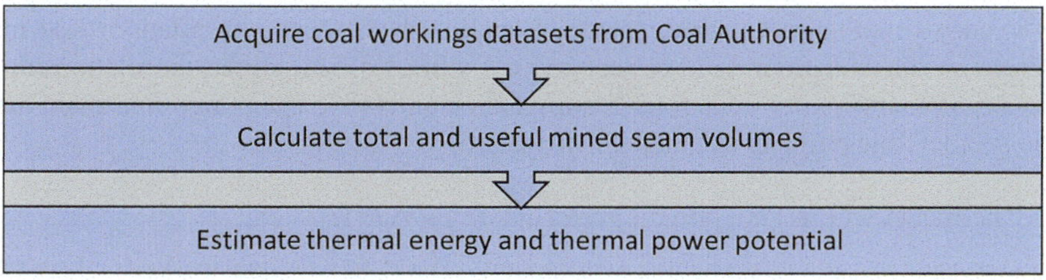

Fig. 8 Project's methodology outline

4.1 Estimating the Heat Potential

The data sought from the Coal Authority included datasets including polygons of underground workings and spine roadways, a table of seam names and their codes, in-seam levels and contours, and working dates. Additionally, BGS borehole data were provided along with a digital terrain model (DTM) of Leeds, and these along with the CA data were processed using QGIS.

Firstly, the seam names were added to the workings' polygons to analyse individual seams as separate layers. The area and volume of the polygons were calculated, followed by the extraction of Z-values (or heights) from polygons and the DTM to determine the depth of each polygon. Mean Z-value was extracted from the DTM, while the majority Z-value was extracted from workings' polygons. This may have resulted in some depth inaccuracies, although they are estimated to be insignificant on the study area scale.

Once depth below ground level (bgl) was calculated, the mine water temperature was estimated for each polygon using the geothermal gradient provided by [8] (Eq. 2).

$$\text{minewatertemperature}(^{\circ}\text{C}) \approx (\text{seamdepthbgl} * 0.036) + 9.5 \tag{2}$$

With seam area, volume, depth bgl, and mine water temperature calculated, the next step in the process was to disregard seams less than 1 m in thickness from the analysis. It was determined that these were unlikely to hold enough volume for efficient water extraction. For further analysis, the remaining seams were subdivided into three groups: seams $\geq$ 1 m, seams $\geq$ 1.5 m, and seams $\geq$ 2 m. Seams thicker than 2 m were likely to have higher thermal potential, with seam thickness acting as a proxy for volume. The overall thermal energy available for extraction in joules was calculated over a range of ΔT values (1°, 2°, and 5°) using Eq. 3:

$$\begin{aligned}\text{ThermalEnergy}\,(\text{J}) = {} & \text{volume}\left(\text{m}^3\right) * \text{density}\left(\text{kgm}^{-3}\right) \\ & * \text{specificheatcapacity}\left(\text{Jkg}^{-1}\,\text{K}^{-1}\right) * \Delta T(\text{K})\end{aligned} \tag{3}$$

Finally, to calculate the total thermal power potential of the workings, assumptions need to be made for the Coefficient Of Performance (COP) of heat pump and the efficiency of submersible water pump. Equation 4 shows the calculation used to determine the total ground source heat (in kW), while Eq. 5 highlights the relationship between ground source heat, total heating effect, and COP. The power consumed by the submersible pump is calculated using Eq. 6.

$$\text{Thermal Power(W)} = \text{pumping rate}\left(\text{m}^3\,\text{s}^{-1}\right) * \Delta T(\text{K}) * \text{specific heat capacity}\left(\text{Jkg}^{-1}\,\text{K}^{-1}\right) * \text{density}\left(\text{kgm}^{-3}\right) \quad (4)$$

$$\text{Ground source heat} = \text{total heating effect} * \left(1 - \frac{1}{\text{COP}}\right) \quad (5)$$

$$\text{Power(W)} = \frac{Q\left(\text{m}^3\text{s}^{-1}\right) * h(\text{m}) * g\left(\text{ms}^{-1}\right) * \rho\left(\text{kgm}^{-3}\right)}{\text{pump efficiency}} \quad (6)$$

It was assumed that the pump efficiency was 50% and that the COP was fixed at 1.5 regardless of mine water temperature. This approach underestimates power calculations because the closer the mine water temperature is to the heating delivery temperature, the higher the COP value would be. With these assumptions, calculations were performed to determine overall heating, ground source heating, heat pump power, and water pump power at five pumping rates—200, 500, 1000, 2000, and 3000 m^3/day.

5 Key Findings of QGIS Analysis

5.1 Temperature and Volume of Mine Workings

The average temperature of mine water within the workings was estimated at 18 °C with a highest of 36 °C and a lowest of 10 °C. The depth of workings directly correlates with the temperature shown in Figs. 9, 10 and 11. The deepest and thus highest temperature block of workings has seam thicknesses between 1.4 and 2.3 m and appears in all three figures. They are found in the southeast corner of the study area beneath Knottingley.

Figure 12 shows the amount of reservoir volume present by increasing seam thickness, and that seams between 1 and 1.5 m make up the greatest proportion of the study area's reservoir volume, at 47% of the total volume. Seams less than 1 m, discarded from the total volume, account for 58% of the total volume.

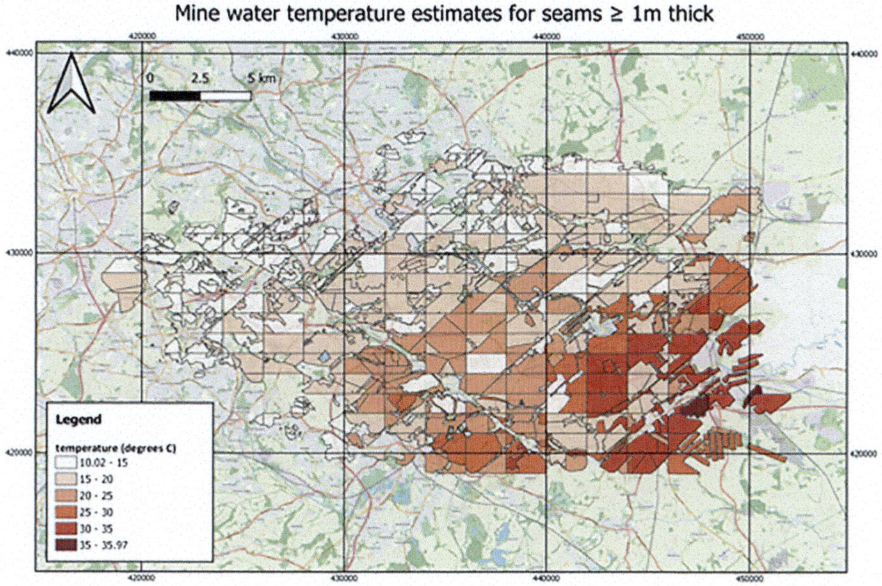

Fig. 9 Estimated temperature of flooded mine workings in seams greater than or equal to one metre

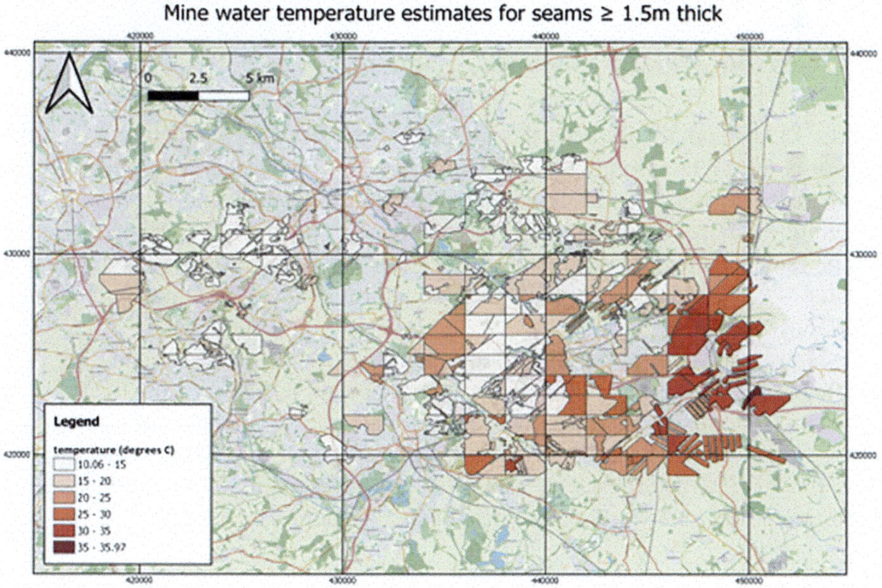

Fig. 10 Estimated temperature of flooded mine workings in seams greater than or equal to one and a half metres

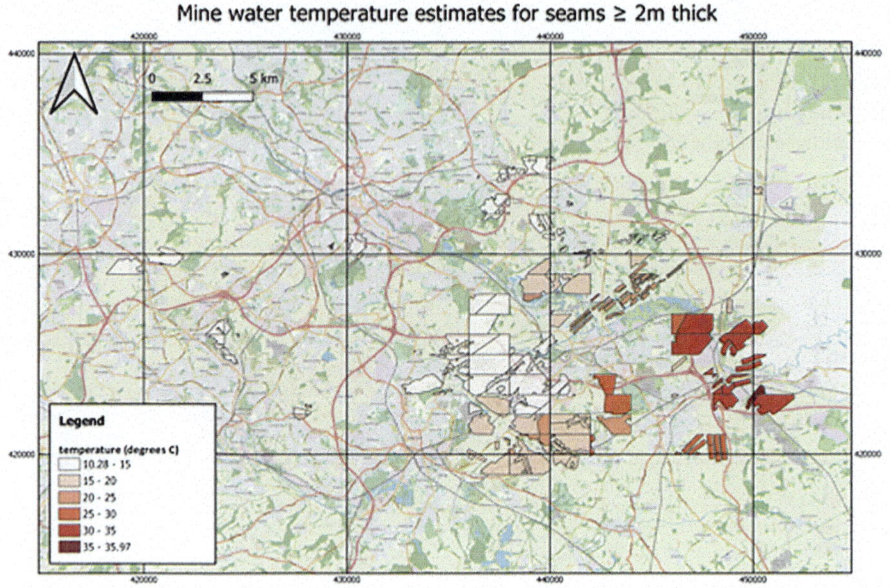

Fig. 11 Estimated temperature of flooded mine workings in seams greater than or equal to two metres

5.2 *Thermal Energy Potential*

Figures 13, 14 and 15 illustrate the spread of thermal energy potential across the study area in MWh, assuming the seams have a void percentage of 70% and 5° of heat are extracted. These are the maximum calculated values and illustrate how—unlike mine water temperature—the thermal energy potential of seams is linked to volume rather than seam depth. The blocks of workings with the most significant potential—17,800 MWh—are found beneath Normanton, approximately the centre of the study area.

In contrast to the highest temperature workings at 700 m depth, they are only 63–100 m bgl. In addition, these blocks are both over 2 m thick. As the thickness of the seams increases, the average thermal potential energy also increases. Workings in Fig. 13 have an average of 1,403 MWh, compared with 1,731 MWh (Fig. 14) and 1817 MWh (Fig. 15). These low values of thermal potential highlight that a large proportion of the total workings have a relatively small overall volume.

5.3 *Power and Pumping Potential*

Figure 16 illustrates the amount of power generated at various pumping rates using a heat pump with a COP of 1.5 to extract 5° of heat from the pumped mine water. The

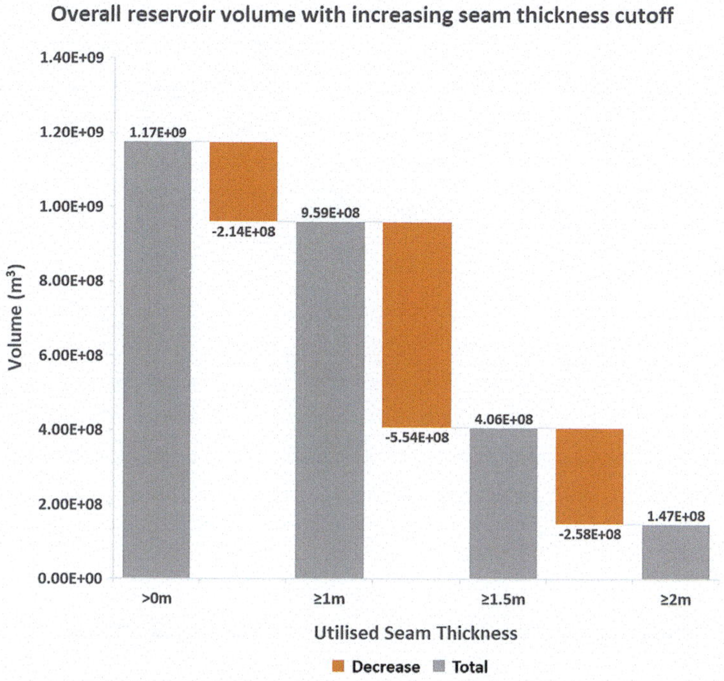

Fig. 12 Waterfall chart showing reservoir volume decrease with increasing seam thickness cut-off. Decrease bars show the amount of reservoir volume that exists between cut-off values

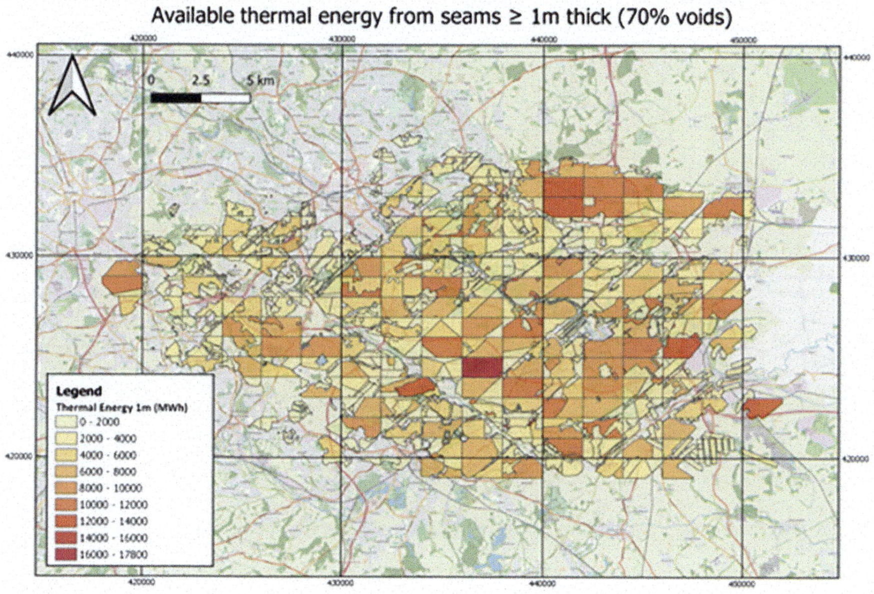

Fig. 13 Thermal energy potential of seams thicker than 1 m expressed in MWh

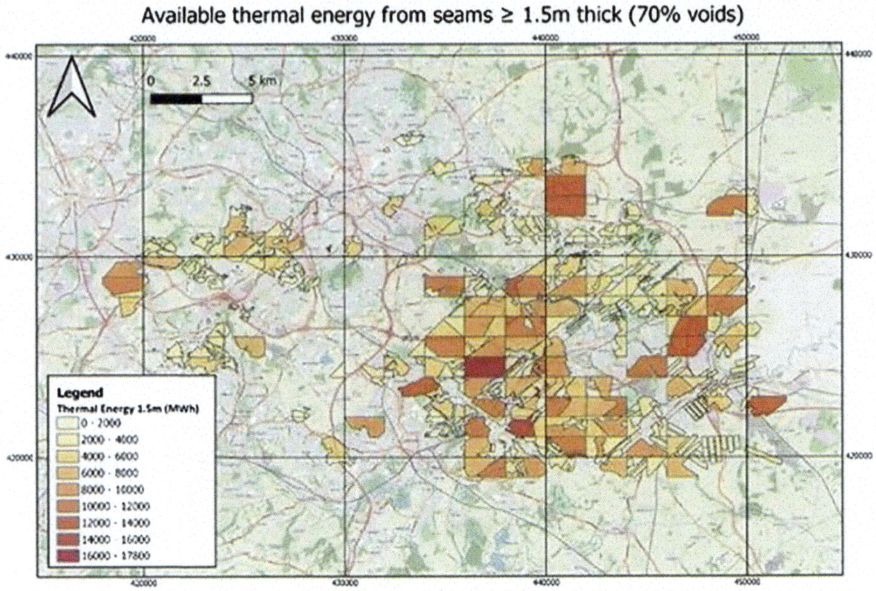

Fig. 14 Thermal energy potential of seams thicker than 1.5 m expressed in MWh

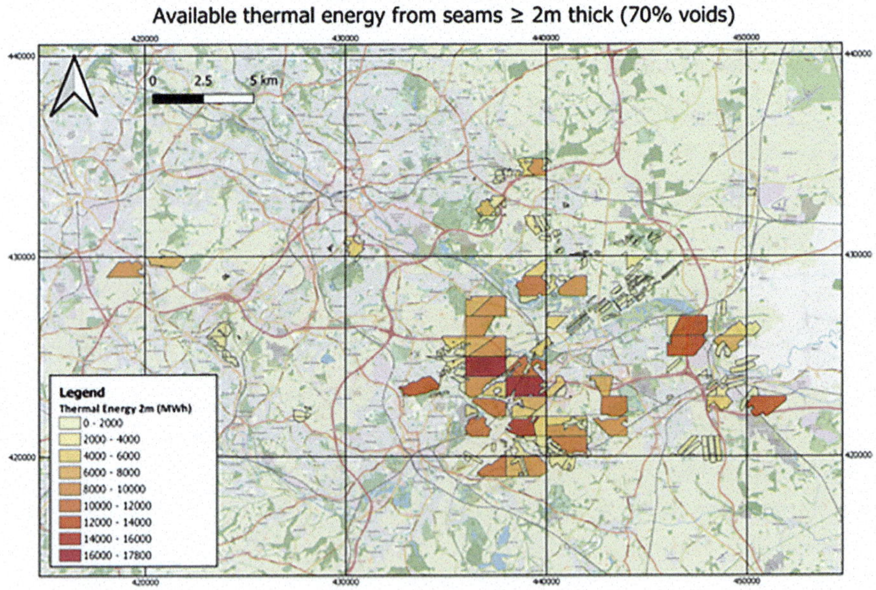

Fig. 15 Thermal energy potential of seams thicker than 2 m expressed in MWh

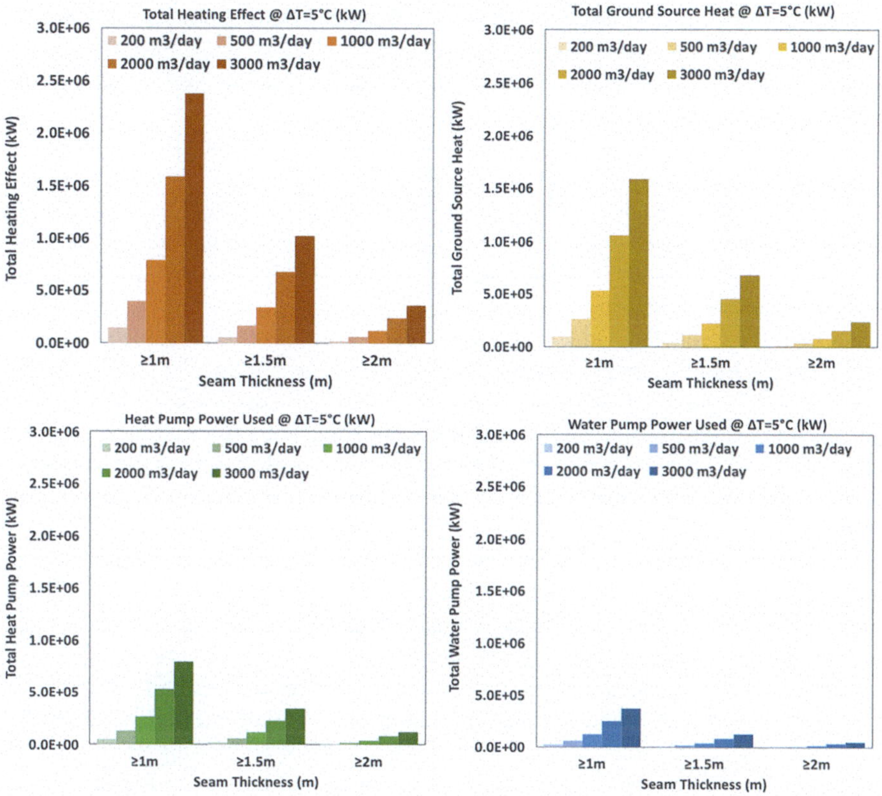

Fig. 16 Top: Thermal power generated versus thermal power extracted from mine workings. Bottom: Power consumed by heat pump and water pump. Total heating effect = ground source heat + heat pump power consumed

figure demonstrates that the ground source heating power generated is significantly greater than the power consumed by the water pump, thereby increasing the overall net power produced.

6 Conclusion

Underground spaces represent a critical asset for cities worldwide, offering immense potential to contribute to long-term sustainability and resilience. This research underscores the importance of approaching underground development comprehensively within the urban ecosystem, taking into account spatial organisation, engineering challenges, policy frameworks, regulations, and social considerations. It demonstrates how abandoned mine workings can be repurposed to serve as sources of sustainable energy, particularly for addressing heating needs in low-income households and combating fuel poverty. By leveraging such resources, cities can make significant strides towards decarbonising their energy systems and mitigating the impacts of the climate crisis.

In the context of Leeds, where 17% of households experienced fuel poverty in 2019, according to Leeds Observatory, and with the urgent need to decarbonise space heating, the evaluation of mine water geothermal schemes becomes increasingly critical. The study reveals that Leeds possesses significant potential for implementing such schemes, particularly in areas with high volumes of abandoned mine workings, such as the southeast region.

Background research conducted on the study area indicates favourable geological conditions for mine water geothermal schemes, with the presence of coal measures characterised by moderately strong sandstones and mud siltstones. Yorkshire, as a whole, exhibits high average geothermal gradients, suggesting substantial geothermal potential. Historical coal mining practices in the region further support the feasibility of such schemes, with most coal workings following room and pillar layouts for centuries.

The project utilised Coal Authority datasets to estimate the potential thermal energy and power that could be extracted from abandoned mine workings, under the assumption that they were fully flooded. While variations in estimates exist, the findings offer valuable insights into the potential magnitude of thermal energy extraction and provide an upper limit to what can be achieved within Leeds.

Overall, this research underscores the importance of exploring innovative solutions for sustainable urban development, particularly in addressing energy needs and combating fuel poverty. By harnessing the untapped potential of underground resources, cities like Leeds can pave the way towards a more resilient, sustainable, and equitable future.

7 Areas of Further Research

Utilising existing geological resources, such as mine workings, offers a significant advantage, particularly in addressing heating needs for low-income households and alleviating fuel poverty. However, challenges arise from the uneven spatial distribution of mine workings across the Greater Leeds Area and data inadequacies.

Sustainability considerations lack comprehensive data on pumping drawdown, recharge rates of mine water levels, and connections between blocks of workings. Despite these limitations, the energy potential estimates offer valuable insights into previously unquantified resources. Refinement of these results through further research and data collection can better inform future mine water geothermal projects.

Acknowledgements The authors would like to acknowledge Coal Authority for sharing the mine working plans, Dr. Sandra Piazolo from the University of Leeds and Dr. Timothy I. Kearsey for publishing this work with the permission of the Executive Director of British Geological Survey.

References

1. Athresh AP, Al-Habaibeh A, Parker K (2016) The design and evaluation of an open loop ground source heat pump operating in an ochre-rich coal mine water environment. Int J Coal Geol 164:69–76
2. Andrews BJ, Cumberpatch ZA, Shipton ZK, Lord R (2020) Collapse processes in abandoned pillar and stall coal mines: implications for shallow mine geothermal energy. Geothermics 88:101904
3. Banks D, Athresh A, Al-Habaibeh A, Burnside N (2019) Water from abandoned mines as a heat source: practical experiences of open- and closed-loop strategies, United Kingdom. Sustain Water Resour Manag 5(1):29–50
4. Banks D, Athresh A, Al-Habaibeh A, Burnside N (2017) Water from abandoned mines as a heat source: practical experiences of open and closed-loop strategies, United Kingdom. Sustain Water Resour Manag 5:29–50. https://doi.org/10.1007/s40899-017-0094-7
5. Benardos A, Athanassiadis I, Katsoulakos N (2014) Modern earth sheltered constructions: a paradigm of green engineering. Tunn Undergr Space Technol 41:46–52
6. British Geological Survey (2021) The BGS lexicon of named rock units—result details: Pennine lower coal measures formation. https://webapps.bgs.ac.uk/lexicon/lexicon.cfm?pub=PLCM. Accessed 21 June 2021
7. Burnside NM, Banks D, Boyce AJ (2016) Sustainability of thermal energy production at the flooded mine workings of the former Caphouse Colliery, Yorkshire, United Kingdom. Int J Coal Geol 164:85–91
8. Farr G, Busby J, Wyatt L, Crooks J, Schofield DI, Holden A (2020) The temperature of Britain's coalfields. Q J Eng Geol Hydrogeol, qjegh2020-109
9. GeoERA (2021) Fact sheet 00—overview of shallow geothermal energy systems in urban environments. https://geoera.eu/projects/muse3/fact-sheets-of-shallow-geothermal-energy-concepts. Accessed 6 Mar 2021
10. HM Government (2020) Powering our net-zero future. Energy white paper, CP 337. HM Government. https://assets.publishing.service.gov.uk/government/uploads/system/uploads/attachment_data/file/945899/201216_BEIS_EWP_Command_Paper_Accessible.pdf
11. Jardón S, Ordóñez A, Álvarez R, Cienfuegos P, Loredo J (2013) Mine water for energy and water supply in the central coal basin of Asturias (Spain). Mine Water Environ 32(2):139–151
12. Monaghan AA, Starcher V, Barron HF, Shorter K, Walker-Verkuil K, Elsome J, Kearsey T, Arkley S, Hannis S, Callaghan E (2022) Drilling into mines for heat: geological synthesis of the UK Geoenergy Observatory in Glasgow and implications for mine water heat resources. Q J Eng Geol Hydrogeol 55(1):qjegh2021–033. https://doi.org/10.1144/qjegh2021-033
13. Ng C, Paraskevopoulou C, Shaw N (2019) Stability analysis of mineshafts used for minewater heat recovery in the UK. Geotech Geol Eng 37:5245–5268. https://doi.org/10.1007/s10706-019-00978-y
14. Paraskevopoulou C, Cornaro A, Admiraal H, Paraskevopoulou A (2019) Underground space and urban sustainability: an integrated approach to the city of the future. In: Proceedings of the international conference on changing cities IV: spatial, design, landscape & socio-economic dimensions. 4th international conference on "CHANGING CITIES: spatial, design, landscape & socio-economic dimensions, 24–29 Jun 2019, Chania, Greece. University of Thessaly. ISBN 978-960-99226-9-2
15. Paraskevopoulou A, Cornaro A, Paraskevopoulou C (2022) Underground space and street art towards resilient urban environments. In: Proceedings of the international conference on changing cities IV: "Making our Cities Resilient in times of Pandemics". June 2022, Corfu, Greece
16. Peralta Ramos E, Breede K, Falcone G (2015) Geothermal heat recovery from abandoned mines: a systematic review of projects implemented worldwide and a methodology for screening new projects. Environ Earth Sci 73(11):6783–6795

17. Sattler T, Paraskevopoulou C (2019) Implications on characterizing the extremely weak Sherwood sandstone: case of slope stability analysis using SRF at two oak quarry in the UK. Geotech Geol Eng 37:1897–1918 (2019). https://doi.org/10.1007/s10706-018-0732-3
18. Verhoeven R, Willems E, Harcouët-Menou V, De Boever E, Hiddes L, Veld POT, Demollin E (2014) Minewater 2.0 project in Heerlen the Netherlands: transformation of a geothermal mine water pilot project into a full scale hybrid sustainable energy infrastructure for heating and cooling. Energy Procedia. Elsevier Ltd, pp 58–67
19. Wray DA (1948) British regional geology: the Pennines and adjacent areas, 2nd ed. His Majesty's Stationary Office, London. http://hdl.handle.net/2027/mdp.39015010048166. Accessed 21 Jun 2021
20. Yiftachel O (2016) The aleph—Jerusalem as critical learning. City 20(3):483–494. https://doi.org/10.1080/13604813.2016.1166702
21. United Nations Summit (2015) 70th Session of the general assembly: report on sustainable development, New York, USA

BiodiverCITY: Urban Soil Life as Vital Underground Infrastructure for Urban Climate Resilience

Joyce van den Berg and Alexandra Pixley

Abstract Climate change significantly impacts urban landscapes, necessitating adaptive urban planning strategies. Soil health and biodiversity are integral to climate resilience as they support ecosystems and green infrastructure that mitigate these effects. However, the invisible world of urban soil life receives little to no attention in urban planning and many knowledge gaps persist. Urban soil biodiversity supports nutrient cycles, water infiltration, and carbon sequestration, essential for resilient urban ecosystems. But urban soils face many challenges from compaction, sealing, pollution, and invasive species, impacting soil health and biodiversity. Effective urban planning must address these issues, integrating soil health into climate adaptation strategies. Amsterdam's BiodiverCITY research project bridges the knowledge gap between science and urban design to highlight the critical role of urban soil biodiversity in ensuring cities remain livable and climate resilient in the future. It has effectively put soil biodiversity on the map for policymakers and urban planners in the Netherlands and beyond.

1 Climate Change and Urban Soils

Climate change presents multifaceted challenges for urban environments, requiring innovative, interdisciplinary, and adaptive strategies in urban planning. Global temperatures have increased by approximately 1.1 °C since pre-industrial times, with predictions of further increase and disastrous consequences if the current emission trend continues [1]. The frequency of extreme weather events, such as heatwaves, heavy rainfall, and droughts, is also on the rise, impacting urban infrastructure and exacerbating biodiversity loss. Urban policymakers in several parts of the world have

J. van den Berg (✉)
Karel Doormanstraat 149-2, 1055VE Amsterdam, The Netherlands
e-mail: Joyce.van.den.Berg@amsterdam.nl

A. Pixley
Nieuwe Amstelstraat 63, 1011PL Amsterdam, The Netherlands

© The Author(s), under exclusive license to Springer Nature Singapore Pte Ltd. 2025
M. Agrawal (ed.), *Underground Spaces for Climate Resilience and Sustainability*,
Advances in 21st Century Human Settlements,
https://doi.org/10.1007/978-981-96-0547-7_7

set ambitious goals to reduce greenhouse gas emissions, enhance green infrastructure, and promote sustainable urban practices.

Early eukaryotic life forms, such as fungi, algae, and protozoa, played a pivotal role in shaping biodiversity that supports modern terrestrial ecosystems. Today, the challenge of soil degradation is a critical global issue, as attested by FAO's Soil Portal [2] and the UN's Global Soil Partnership (GSP) [2] Loss of topsoil impacts ecological functions, exacerbates global warming, and demands urgent attention—not only in agriculture but also in urban planning. Soil health is integral to climate resilience, supporting urban ecosystems that mitigate the effects of climate change in urban areas.

Amsterdam's economic and population growth is putting pressure on public space and its underground space, with the city's population projected to increase by 21% by 2050 [3]. Furthermore, ambitious climate and sustainability goals require large-scale upgrading of existing infrastructures such as energy and heating systems, digital networks, and sewage systems. Together, this means an increasing need for comprehensive and integrated planning of public spaces and full commitment to greening the city to enhance soil health to combat the urban heat island effect, manage rainwater effectively, and protect urban biodiversity. Additionally, inadequate underground data often results in damage to existing subsoil and underlying infrastructure, emphasizing the need for comprehensive soil management and precise mapping of underground networks.

Efficient soil management practices can mitigate climate pressures by improving water infiltration, reducing floods, and supporting biodiversity. Urban planners must prioritize soil health to create resilient, sustainable cities that can adapt to the challenges posed by climate change and urbanization.

2 BiodiverCITY: A Matter of Vital Soil

In 2021, the City of Amsterdam launched the BiodiverCITY research project to explore urban soil biodiversity, resulting in the publication of *BiodiverCITY_A Matter of Vital Soil!*. This award-winning initiative highlights the important role of soil in urban systems and makes the often invisible world of soil life visible. By synthesizing scientific knowledge on biodiversity and urban ecology with practical insights from urban design and maintenance processes, BiodiverCITY offers a comprehensive overview of the knowns and unknowns of urban soil biodiversity. BiodiverCITY is aimed at urban professionals and the general public in the Netherlands and beyond, especially calling on designers and policymakers to take action. The project's compelling visual language brings the microscopic world of soil biodiversity to life, inviting readers to explore the essential ecosystems beneath their feet.

BiodiverCITY is the result of an interdisciplinary collaboration among scientists, urban designers, landscape architects, policymakers, and artists. This collaborative approach bridges professional languages and perspectives, enhancing the

Fig. 1 *BiodiverCITY_A Matter of Vital Soil!* is a research and publication that has been well received in the Netherlands and abroad. The printed edition has sold out, but it is available online for free in English and Dutch at https://openresearch.amsterdam/nl/page/71091/biodivercity. Image by Alexandra Pixley

project's depth, reach, and innovative potential. Through a combination of long-form essays, scientific articles, concise explanations, and visually intricate infographics, the project conveys the complexity and importance of soil biodiversity for the future of urban spaces. By presenting complex information in an accessible and engaging format, BiodiverCITY aims to inspire a diverse audience—from urban specialists to residents—highlighting the critical role of soil biodiversity in urban ecosystems (see Fig. 1).

3 Why Soil Biodiversity?

The top layer of soil, particularly the uppermost meter of the earth's crust, is crucial for sustaining life on Earth. Without healthy soils, terrestrial life cannot exist. A single cup of soil contains approximately 200 billion bacteria, 100,000 m of fungal hyphae, 20 million protozoa, 100,000 nematodes, 50,000 arthropods, and nearly one earthworm [4]. A teaspoon of soil contains more living organisms than people on Earth, illustrating the biodiversity richness of the soil. Soil biodiversity is the foundation of terrestrial life providing critical functions such as nutrient cycling, decomposition, soil formation, rainwater retention and filtration, climate regulation, carbon sequestration, and the fertile ground for plants to establish roots and for several creatures, like insects and reptiles, to thrive. Poor soils with low biodiversity will underperform in all these areas, ultimately affecting the health and diversity of plants and animals that can live in and upon that soil.

Furthermore, healthy soils play a crucial role in mitigating the effects of climate change. In urban settings, these soils enhance water infiltration and storage, reduce the urban heat island effect, and store significant amounts of carbon. Addressing soil health is thus essential for managing the impacts of extreme weather events, such as heavy rainfall and droughts, and amplifying urban resilience. As soil health is directly related to plant health, urban soils also impact the quality of greenery in urban spaces. Green infrastructure is critical for climate resilience but also provides a range of visual, phychological, and social benefits to urban dwellers.

Soil health is determined by a combination of physical, chemical, and biological properties. Physical properties relate to the soil's structure and its capacity to retain water. Chemical properties focus on soil's ability to retain nutrients and regulate pH levels. Historically, soil health was assessed solely on these two properties. However, understanding the biological properties—particularly soil life—is crucial for a comprehensive assessment of soil health. Soil organisms play vital roles in loosening soil, facilitating water infiltration, adding humus, suppressing diseases and pests, and neutralizing contaminants.

Healthy soil reduces erosion, stores large quantities of CO_2, and supports resilient ecosystems. By studying soil biodiversity, we can better understand the complex interactions that sustain soil health and, by extension, the health of urban ecosystems (see Fig. 2).

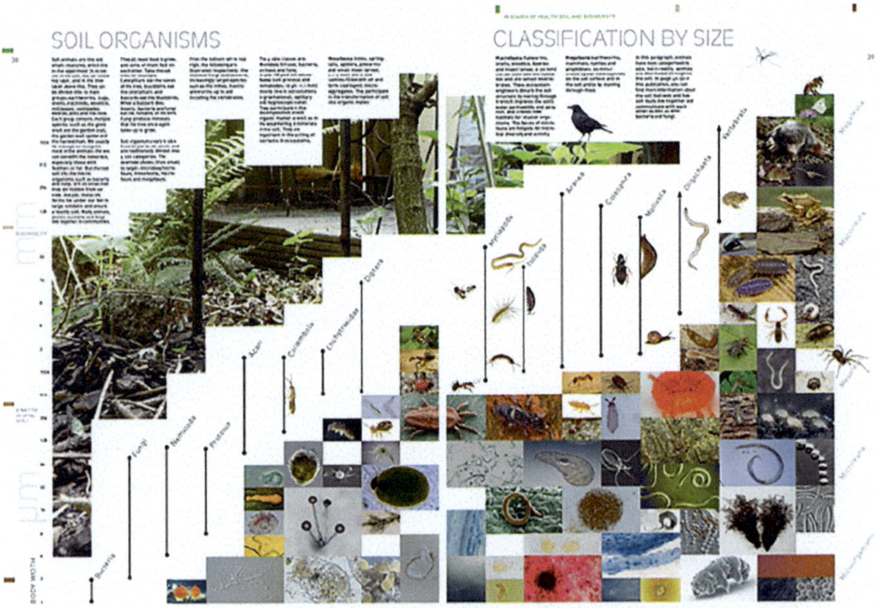

Fig. 2 Excerpt from *BiodiverCITY_A Matter of Vital Soil!* publication shows an overview of the different kinds of soil organisms that can be found in urban soils, categorized by size and taxonomy. *Source* Berg, Joyce van den (et al.). 2021. *BiodiverCITY_A Matter of Vital Soil!*. Rotterdam: nai010 publishers

4 The State of Our Urban Soils

Urban soils are shaped by natural processes and human activities, with cities consisting of homes, streets, public spaces, and various infrastructure elements. Urbanization exerts a heavy toll on soil health. Urban soils face considerable pressure from compaction, hardening, excavation or soil removal, pollution, and the introduction of invasive species. Despite these challenges, urban soils can harbor significant biodiversity. For instance, studies [5] have found as much biodiversity in the soil of New York's Central Park as in a tropical rainforest.

The expansion of cities, driven by technological advancements and population growth, has significantly altered soil landscapes. Urban infrastructure development often disrupts rich soils and habitats, leading to a decline in soil health and biodiversity. Historically, buildings in the Netherlands were constructed on naturally higher ground. Over time, land for cities was drained, canals were dug, and embankments were built, enabling construction in various areas. Technological advancements in the late nineteenth century further transformed urban landscapes, leading to extensive underground networks of water mains, sewerage systems, and utility cables. Due to the unstable river bedding upon which most of Amsterdam has been constructed, land subsidence is a huge issue. For this reason, houses are built on piles. However, street levels and public spaces need to be periodically raised to compensate for subsidence, causing significant soil disturbance and altering the composition of the topsoil. As cities continue to grow, the pressures on urban soils will only increase. Effective soil management practices, informed by both historical and modern insights, are essential for maintaining healthy urban ecosystems. Modern urban planning must reconcile the need for infrastructure development with the conservation of soil health.

Urban soils face unique challenges that require targeted measures to address declining biodiversity and habitat destruction. Effective urban planning must consider these challenges and implement strategies to protect and enhance soil health. Several key challenges can be identified.

Compaction

Compaction resulting from buildings and underground structures poses a significant threat to urban soils. Compacted soil suffers from reduced oxygen levels and diminished water retention capacity, negatively affecting soil life and tree roots.

Sealing

Moreover, the sealing of soil surfaces by buildings, roads, and pavements hampers rainwater infiltration, causing water accumulation in low-lying areas and elevating the risk of flooding. Sealed soils exhibit reduced biodiversity and lack the interactions with surrounding ecosystems characteristic of healthy, open soils.

Pollution

Urban soils are often contaminated by pollution from industrial activities, construction sites, and the use of pesticides, herbicides, and chemical fertilizers in public green spaces. Pollution disrupts soil ecosystems, leading to decreased biodiversity and compromised soil health.

Invasive Species

Additionally, invasive species disrupt ecological balance and outcompete native species, exacerbating the decline in soil biodiversity. To effectively manage urban soils, it is imperative to address pollution and invasive species by promoting practices that restore soil health and support native biodiversity. This includes reducing the use of harmful chemicals and implementing strategies to control invasive species.

Disturbance from Urban Development

Minimizing soil disturbance during construction activities is essential for preserving the health of urban soils, particularly for fungal networks, whose mycelial structures are sensitive to disruption. To achieve this, careful planning of construction activities is paramount, along with the use of light and small equipment to minimize soil disruption. Additionally, preserving existing soil structures wherever possible can help maintain soil health and support urban ecosystems (see Fig. 3).

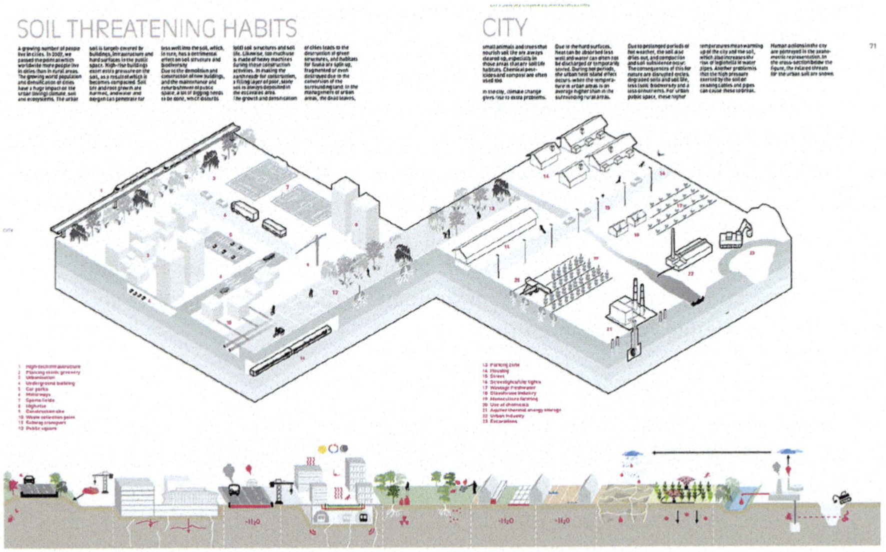

Fig. 3 Urban soils are under many pressures from different industries, buildings, mobility, interventions and management habits. *Source* Berg, Joyce van den (et al.). 2021. *BiodiverCITY_A Matter of Vital Soil!*. Rotterdam: nai010 publishers

5 Soil in Urban Design: Amsterdam's Integral Design Method for Public Spaces

Urban planners need to include soil health as an integral part of climate adaptation strategies. This involves implementing practices that enhance soil permeability, promote water storage, and support microbial life—to foster resilient urban environments that can withstand the challenges posed by climate change. Design strategies should include measures to protect existing soil biodiversity, promote native species, and restore degraded soils. These efforts support healthy urban ecosystems and improve the overall quality of urban life.

Amsterdam's Integral Design Method Public Space (Dutch acronym is IOOR) prioritizes soil and subsurface space as crucial components of urban design and advocates a fundamentally different approach to design that supports resilient urban ecosystems. By emphasizing early interdisciplinary collaboration, involving all stakeholders in the development of public space, and designing the underground and subsurface in conjunction with the public space, this method offers a comprehensive approach that can serve as a model for other cities looking to integrate soil biodiversity into their planning processes.

The city of Amsterdam has pioneered the IOOR method, highlighting the essential role of soil in urban infrastructure. Faced with ecological crises such as climate change, biodiversity loss, and pollution, Amsterdam recognizes the necessity of understanding and managing urban soil. Planned transformations, including the installation of new energy systems, replacement of aging sewage systems, and more space for green areas, will significantly impact soil health. A deliberate urban soil strategy is crucial, especially considering that each of the city's 4,800 streets is projected to undergo construction multiple times by 2040.

Integrating soil health into urban planning is not just a technical necessity but a cornerstone for the future liveability and climate resilience of cities. Healthy soils support plant life and larger ecological cycles. Urban soil strategies must therefore address both the immediate needs of infrastructure projects and the long-term sustainability of urban ecosystems.

5.1 Amstel-Stad Case Study, Amsterdam

Amstel-Stad, the project that birthed IOOR method, presented a unique challenge as a transformation area. Existing streets, with fixed widths, posed the challenge of accommodating additional cables and pipes alongside existing infrastructure. However, at many locations, space was severely limited. Complicating matters further, the soil in Amstel-Stad exhibited a highly heterogeneous structure and complex hydrology. Traditional design processes would have struggled to accommodate the diverse range of necessary infrastructure and "infra-nature" (nature-based infrastructure), including adaptations for climate resilience, new energy systems, heavier electrical and data cables, sewage systems, and other sustainability goals.

5.2 Interdisciplinary, Integral, Interscalar, and Research-Based

The Integral Design Method Public Space represents a paradigm shift in urban planning methodologies. It encompasses an extensive research phase, mapping detailed climate impact data and analyzing spatial requirements in depth. Moreover, it emphasizes interdisciplinary collaboration among various urban stakeholders and public space asset owners throughout the entire process.

One of its innovative aspects is the introduction of an "in-between scale," which bridges the gap between traditional urban development projects and city-wide masterplans and policies. This approach allows for more nuanced design and analysis, facilitating a precise definition of sustainability objectives through the use of the "ambitionweb tool" and performance goals. These frameworks enable the negotiation of performance goals and ambitions at different levels, from entire neighborhoods to individual streets, before diving into specific design solutions.

Another key aspect is the method's growing catalogue of modular and multifunctional spatial measures, which can be tailored and adapted to suit the specific context of each street profile. This flexibility ensures that solutions are not only effective but also responsive to the diverse needs and challenges of urban environments.

The measures and resulting details in the Handbook are divided according to seven area themes—Living Environment, Water, Flora and Fauna, Energy, Soil and Subsurface, Mobility, Materials—and contain the smart stacking of different functions, like trees with water retention or deeply placed transport pipes with vegetation or trees growing on top. The catalogue shows the high-potential combinations between resulting details from the different themes, as illustrated in Fig. 4.

Following the steps of the integral design method public space, as outlined in more detail in Fig. 5, will result in designs that integrate all the different aspects, disciplines necessary to ensure future-proof cities. Thus, public space designs and strategies are developed that incorporate soil and subsurface space to achieve climate adaptation and sustainability goals.

6 Strategies and Measures for Enhancing Soil Health

In the context of BiodiverCITY, the catalogue of measures and resulting details of the Integral Design Method Public Space was expanded with measures specifically to support and improve soil health and biodiversity. The developed biodiversity measures fall into four categories: Ecological Matrix, Water, Soil, and Management.

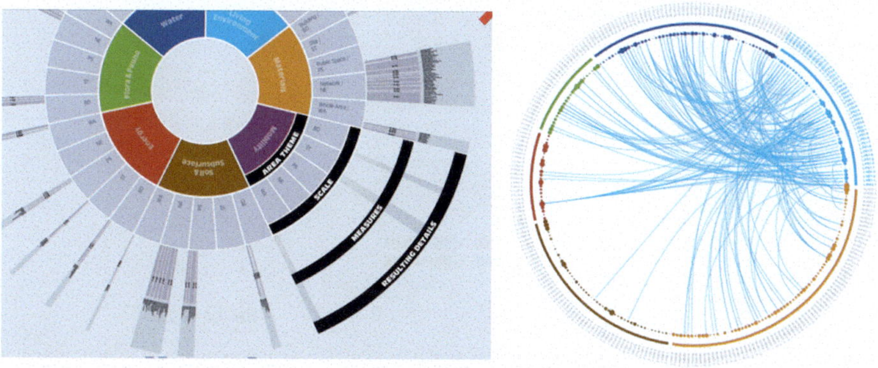

Fig. 4 Per area theme, different measures and resulting details have been developed and bundled in the IOOR Handbook. These two excerpts from the Handbook illustrate how each measure is stacking functions and connects to different themes, illustrating the fundamentally integral and multifunctional nature of the IOOR approach and the design solutions it proposes. *Source* Berg, Joyce van den (et al.). 2019. *Integral Design Method Public Space, Vol. II—Handbook*. Amsterdam: City of Amsterdam

6.1 Spatial Measures

Measures to improve soil health include promoting soil connectivity, reducing soil disturbance, and enhancing soil biodiversity through ecological practices. Strategies involve soil transplantation, reducing hard impermeable surfaces, and supporting the growth of indigenous plants. These efforts create a more resilient urban ecosystem capable of adapting to changing environmental conditions.

Effective water management practices, such as increasing soil porosity and implementing water storage solutions, are essential for urban resilience. These measures help manage extreme weather events and support healthy soil ecosystems. Other strategies include designing public spaces that promote water infiltration, creating buffer zones, and implementing green infrastructure solutions.

When it comes to management, implementing simple measures can yield significant benefits. For instance, refraining from removing leaf litter from green areas can have a profound impact. Allowing dead leaves to naturally decompose and enrich the soil fosters a healthy soil food web. This approach promotes biodiversity, enhances soil structure, and contributes to overall soil health. By embracing such practices, urban environments can thrive while minimizing human intervention (see Fig. 6).

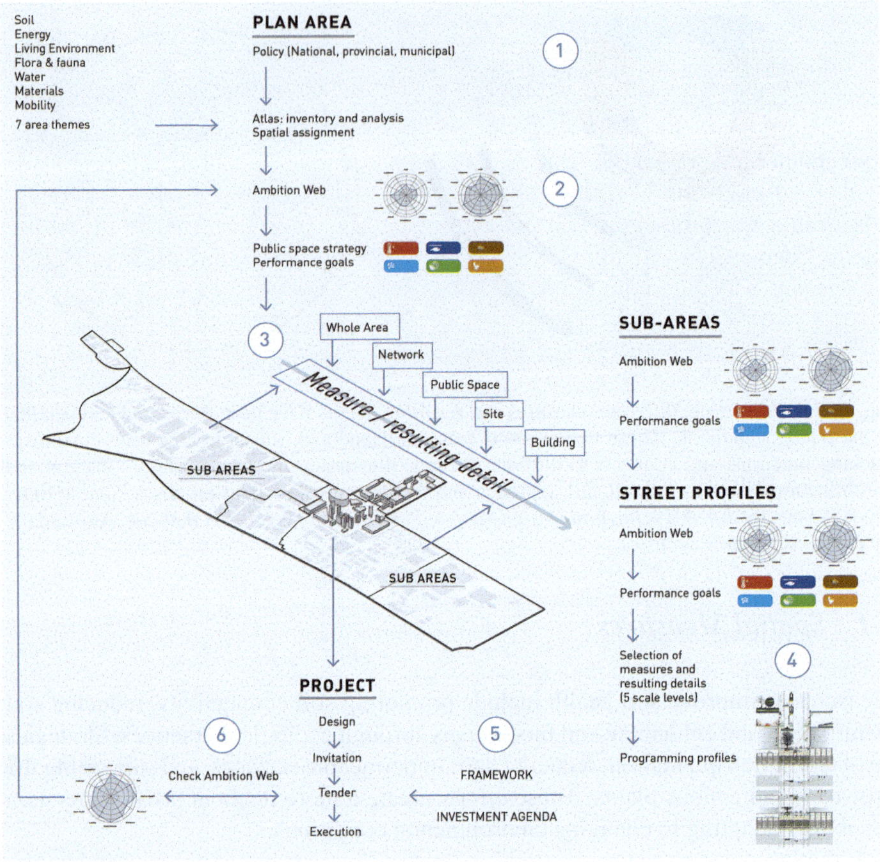

Fig. 5 The 6 steps of the Integral Design Method Public Space. *Source* Berg, Joyce van den (et al.). 2019. *Integral Design Method Public Space, Vol. I—Case Study Amstel-Stad, Amsterdam.* Amsterdam: City of Amsterdam

7 Soil Futures

BiodiverCITY was just the first exploration of the importance and potential of paying attention to life underneath the surface in our urban spaces. It uniquely and effectively puts soil biodiversity onto the agenda for cities across the Netherlands and abroad, especially as a criterion for ensuring the effectiveness of climate adaptation measures—from healthy trees to water retention to a biodiverse urban ecology and a pleasant living environment.

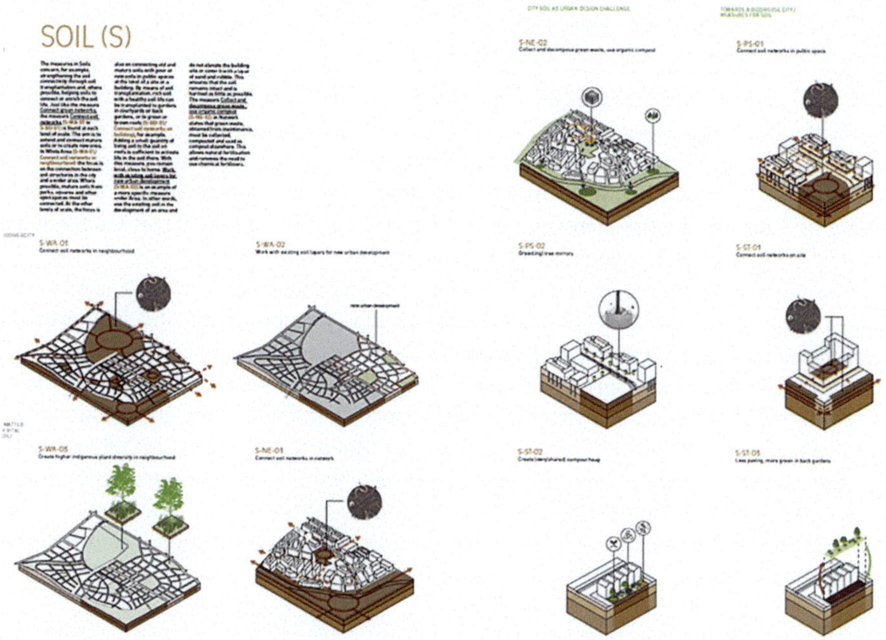

Fig. 6 In BiodiverCITY, measures for Soil were developed to be added to the library of measures as found in the *Integral Design Method Public Space, Vol. II—Handbook*. *Source* Berg, Joyce van den (et al.). 2021. *BiodiverCITY_A Matter of Vital Soil!*. Rotterdam: nai010 publishers

7.1 Policy Frameworks and Legal Protection

The policy and research priorities outlined in BiodiverCITY are guiding the development of new policies, practices, and initiatives aimed at integrating soil biodiversity into urban planning and public space management. These efforts are crucial for equipping cities to address the challenges of the twenty-first century and comply with national and international climate goals and legislation. Establishing clear policy objectives for soil biodiversity is essential to ensure its protection and enhancement. By integrating scientific knowledge into urban planning and legal frameworks, policymakers can promote practices that support soil conservation and enhance resilience. Interdisciplinary collaboration, backed by strong policy support and public engagement, will be vital for unlocking the full potential of urban soils in creating healthy, vibrant cities for generations to come.

7.2 Education and Awareness

In recent years, the Netherlands has seen a surge in urban soil initiatives, as depicted in Fig. 7. Raising awareness about the importance of soil biodiversity through education and citizen science initiatives has become paramount. Engaging the public in soil health projects can deepen their understanding and commitment to sustainable urban practices. Educational programs should emphasize the critical role of soil biodiversity in supporting urban ecosystems and climate resilience.

One notable initiative, "Onder Het Maaiveld" (Below the Surface), led by the nature foundation IUCN NL, the National Butterfly Foundation, the Netherlands Institute of Ecology, Wageningen University and Research, and the Centre for Soil Ecology, aims to make the invisible world of soil life more visible to people of all ages. Through initiatives such as a high-quality production documentary about Dutch soil life and a series of children's books, Onder Het Maaiveld educates audiences about the importance of soil biodiversity and encourages them to care and pay attention.

Another promising initiative is the HiddenBiodiversity project, funded nationally through the Dutch Science Agenda. This project focuses on mapping and investigating urban biodiversity that is often invisible or underexamined, such as bacteria,

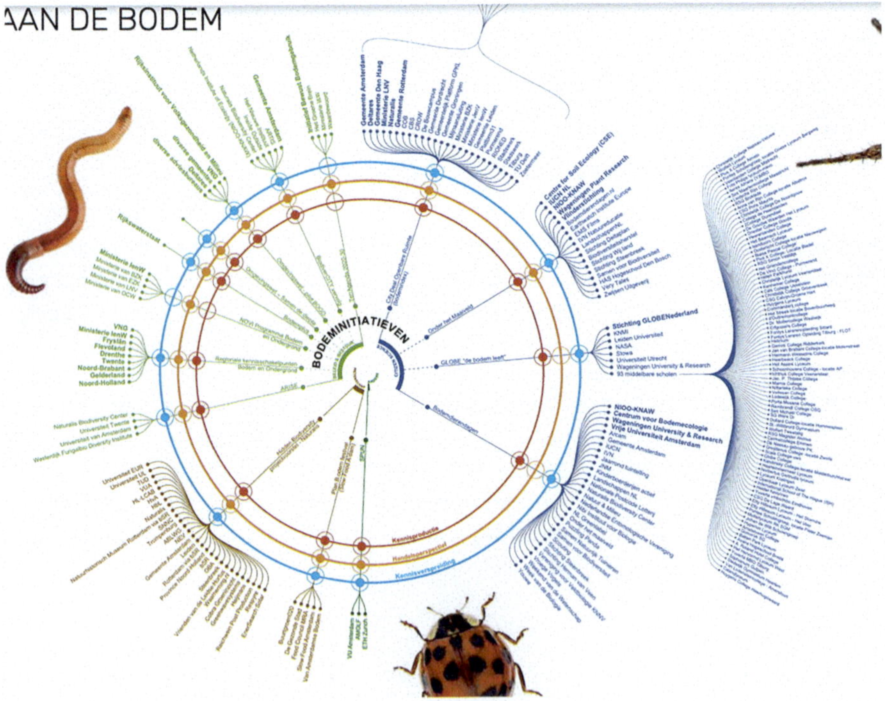

Fig. 7 Initiatives and organizations working on soil life, soil biodiversity, and soil ecology in the Netherlands. *Source* Portfolio Joyce van den Berg

fungi, soil animals, algae, mosses, lichens, and pavement plants in urban spaces. Moreover, the HiddenBiodiversity project places a strong emphasis on engaging and educating urban citizens about this hidden biodiversity.

7.3 Academic Workshop Soil

The Chief Science Office of the City of Amsterdam has launched Academic Workshops, which bring together science, policy, practice, and education. Together, the municipality, academic institutions, and the public delve into urban issues. In continuation of the inquiry into soil and subsurface matters, the Academic Workshop Soil (AWS) was initiated in 2023, focusing on the Integral Design Method Public Space and *BiodiverCITY_A Matter of Vital Soil!*. As its first activity, the AWS organized a lecture series exploring the question: "How does it all come together in the soil of public spaces, now and in the future?" To foster interdisciplinary discussions, four themes were selected: Time, Scale, Depth, and Interaction.

By convening researchers, designers, artists, and the public with diverse perspectives on urban soil, the workshop aims to transcend limited understandings, identify blind spots, and generate new research and design inquiries. Emphasizing integral design and interdisciplinary collaboration, the workshop seeks to contribute to the creation of healthy, livable future-resilient cities. Preliminary findings from themed seminars have been visualized in artworks by Jaehun Park and showcased at the International Architecture Biennale Rotterdam 2024 "Nature of Hope."

The Academic Workshop Soil has raised thought-provoking questions for designers and urban professionals interested in the urban underground, subsurface space, and urban soils. For instance, with uncertainties surrounding geothermal energy production and underground carbon storage, especially regarding microbial life at greater depths, how do we assess risks and establish necessary conditions for installing new infrastructures in the deeper layers of the urban subsurface? How can we effectively balance top-down big data predictions with localized tests and interventions based on local experience and field data? What governance models are suitable for managing the urban subsurface, considering both horizontal (private buildings, public space) and vertical delineations? How can we comprehend and map the scale and interactions of various biotic and abiotic factors and manmade infrastructures in the urban subsurface to optimize underground space utilization and interactions between these networks? Lastly, how can we preserve both our archaeological past and achieve climate resilience in the present urban soil and subsurface management?

These questions underscore the importance of thoughtful and more holistic approaches to urban soil and subsurface issues (see Fig. 8).

Fig. 8 Still images from 3D simulation animation made by artist Jaehun Park for the Academic Workshop Soil. The video animation were on display at the 11th International Architecture Biennale Rotterdam 2024. *Source* Stills from works by Jaehun Park for Academic Workshop Soil

8 Conclusion

What happens underground is going to determine the future of our cities in many ways. To make future-proof cities, soil biodiversity and infra-nature (nature-based infrastructure) can no longer be ignored and must become an inseparable part of city design. In designing public space, more integrated and ground-up design is needed instead of the current way of top-down design that does not take the underground and subsoil into consideration. This article has elaborated on the importance of soil biodiversity as highlighted in Amsterdam's BiodverCITY project that researched and raised awareness on the topic of urban soil. Making the invisible and largely unknown underground world of soil life visible and relatable is a crucial part of centering soil in policy and design. Although there are still many knowledge gaps and unknowns regarding urban soil life and urban soil biodiversity, there are simple yet effective measures that can be taken to enhance the condition of urban soils. When applied as part of a wider integral design—like by applying the Integral Design Method Public Space—these measures are key to ensure the climate resilience of urban areas and combat soil threats like compaction, pollution, and soil impoverishment and sealing.

It is critical to continue research on urban soil and address the many existing knowledge gaps to ensure policy and design can be more effective when it comes to supporting soil biodiversity. This includes getting a better understanding of the interactions between urban soil life and other underground (and aboveground) infrastructures and activities. For this reason the recent initiative of the Academic Workshop Soil is bringing together scientists, designers, policymakers, artists, and the general public, to discover new perspectives at the intersection of different understandings and uses of the ground under our feet. In short, cities need to start making soil a priority in their research, policy-making, and urban design approaches, for the sake of our urban future.

References

1. Intergovernmental Panel on Climate Change (IPCC) (2023) Climate change 2023: synthesis report. Summary for policymakers. https://www.ipcc.ch/report/ar6/syr/downloads/report/IPCC_AR6_SYR_SPM.pdf. Accessed 1 Jul 2024
2. Food and Agriculture Organization of the United Nations (FAO) (2024) FAO soils portal. https://www.fao.org/soils-portal/en/. Accessed 1 Jul 2024
3. Gemeente Amsterdam, Onderzoek en Statistiek (O&S) (2024) Bevolkingsprognose 2023–2050, factsheet April 2023. https://onderzoek.amsterdam.nl/publicatie/bevolkingsprognose-2023-2050. Accessed 2 Jul 2024
4. van den Berg J et al (2021) BiodiverCITY_A Matter of Vital Soil!, Rotterdam: nai010 publishers
5. Korthals G, Rienks F (2021) Bacteria & co support the entire ecosystem. In: van den Berg J et al (eds) BiodiverCITY_A Matter of Vital Soil!. Rotterdam: nai010 Publishers, pp 41–42

Assessing the Value of Qanat System of Yazd in Promoting Urban Climate Resilience

Mahdi Shakibamanesh

Abstract Exploring the qanat system of Yazd, an ancient Persian water management technique, this study delves into its pivotal role in urban climate resilience. Situated on Iran's arid central plateau, Yazd relies on qanats to combat water scarcity, fostering sustainable agriculture and human settlement. These subterranean tunnels, minimizing evaporation, are testaments to ancient engineering prowess, facilitating biodiversity, urban cooling, and soil salinity management while fostering communal cooperation. However, modern challenges like urbanization and groundwater overuse imperil these structures. By integrating contemporary technologies such as IoT and nanotechnology with traditional practices, qanat efficiency and sustainability can be enhanced. Emphasizing community involvement, policy frameworks, and strategic urban planning, this exploration underscores qanats as models for sustainable water management in arid regions. Vital for the ecological and economic stability of Yazd, their preservation and revitalization ensures a steady water flow supporting agriculture, mitigating drought impacts, and water security.

Keywords Qanat System · Urban Climate Resilience · Water Management · Arid Regions · Yazd City

1 Introduction

Yazd, located on Iran's central plateau, faces extreme climatic conditions with an average annual precipitation of about 98 mm, which is significantly lower than Iran's national average of 273 mm and the global average of 990 mm [2, 9]. This low precipitation, coupled with a high mean annual temperature exceeding 22 °C, classifies Yazd within the Aridic-hyperthermic soil moisture and temperature regime [16]. The province covers approximately 74,645 km^2 and experiences an average annual evaporation of about 3193 mm [2].

M. Shakibamanesh (✉)
Faculty of Architecture and Urban Planning, Shahid Beheshti University, Tehran, Iran
e-mail: m.shakibamanesh@alumni.sbu.ac.ir

© The Author(s), under exclusive license to Springer Nature Singapore Pte Ltd. 2025

M. Agrawal (ed.), *Underground Spaces for Climate Resilience and Sustainability*,
Advances in 21st Century Human Settlements,
https://doi.org/10.1007/978-981-96-0547-7_8

Historical evidence indicates that Yazd was once more humid, as suggested by the micromorphological properties of soils which consist of illuviated clay and needle calcite [16]. The current arid conditions have led to the development of subsistence agricultural economies through qanat irrigation, essential in areas with low annual precipitation below 250 mm [13].

The qanat system, an ancient hydrological marvel, originated in Persia around the early 1st millennium BC [5]. This method of water extraction has been crucial for sustaining agriculture and human settlements in arid regions for over 2,500 years. A qanat comprises a nearly horizontal tunnel that collects water from an underground source, typically an alluvial fan, through a mother well sunk to reach the aquifer [10].

There are approximately 22,000 manually constructed qanats in Iran, stretching over 274,000 kms. These qanats collectively produce between 750 and 1,000 cubic metres of water per second, demonstrating their significant contribution to water supply [3]. In Yazd, a city renowned for its extensive qanat network, around 70 qanats, each ranging from 30 to 45 kms in length with mother wells 50–125 m deep, provide essential water resources [6]. The qanat system mitigates water scarcity risks exacerbated by climate change, such as prolonged droughts and irregular rainfall. Its underground design reduces evaporation losses, making it highly efficient in hot climates where water conservation is crucial [18].

2 Historical and Cultural Fabric of Qanats

Qanats are an ingenious solution to water scarcity in arid regions. Historically, these structures enabled human settlements to thrive by providing a consistent water supply for agriculture, drinking, and sanitation. The city of Yazd has long depended on this sophisticated system to sustain its population and agricultural needs in a desert.

The cultural and economic impacts of qanats are profound. They facilitated the development of prosperous communities and urban centres in regions where water was scarce.

Today, their ability to transport water from aquifers using gravity alone makes them both energy-efficient and environmentally friendly [18]. In the face of climate change, which exacerbates water scarcity and drought conditions, the qanat system's resilience is invaluable. Their underground construction minimizes water loss from evaporation and protects the supply from extreme weather events.

By promoting sustainable and equitable water sharing through traditional communal management systems, these ancient structures enhance social cohesion and local capacity to address climate-related challenges [10].

This section delves into the qanat system of Yazd, highlighting its ingenious construction techniques, and crucial role in ancient water management. It also explores the social and cultural dimensions of qanats, underscoring their significance in promoting urban climate resilience in arid regions.

2.1 Construction Techniques and Development

The Qanats, also known as kariz, are an ancient Iranian irrigation system showcasing remarkable engineering and hydrological expertise. These systems consist of deep-water wells interconnected by a network of underground tunnels and canals to form artificial springs. Construction begins with identifying a suitable water source, typically an aquifer at the foothills of mountains. A mother well is dug into the water table, followed by a gently sloping tunnel that channels water to the surface using gravity.

Qanat construction is labour-intensive, requiring precise calculations and skilled labour. Key components include the appearance (where water emerges), the gallery (a horseshoe-shaped underground canal), dry and wet zones, and vertical shafts for soil extraction, ventilation, and maintenance. Deeper wells are positioned closer to the mother well to ensure proper air passage [3].

Qanats balances water extraction with sustainability. The gentle tunnel gradient prevents rapid erosion and collapse, ensuring longevity. The construction techniques passed down through generations of muqannis (qanat builders), involve traditional methods and tools that have changed little over centuries [10].

2.2 The Role of Qanats in Ancient Water Management

In Yazd, the qanat system played a significant role in shaping urban design. The layout of courtyards and narrow streets was strategically developed to manage solar radiation and enhance air circulation, naturally cooling the environment. Traditional architectural elements such as thick mud-brick walls, domed roofs, and wind towers (badgirs) were not only cultural symbols but also practical solutions to mitigate the extreme desert climate. These features reduced the reliance on mechanical cooling systems [23].

Despite their effectiveness, qanat-based agriculture faced significant challenges, particularly from recurrent droughts and increasing population pressures. The water flow in qanats could diminish drastically during periods of low precipitation.

2.3 Social and Cultural Dimensions of Qanats

Qanats in Yazd help promote social equity and communal cooperation. In arid regions where water is a shared and precious resource, the qanat system reflects deep-seated values of collaboration and mutual responsibility. Ownership of water rights has traditionally dictated social hierarchy, with those controlling water held in high esteem. The proximity of one's home to the qanat often determined social status, with upstream locations being more desirable and accessible to wealthier families.

The Ghassem Abad qanat in Yazd exemplifies this communal governance, where the Mirab, or water master, oversees the equitable distribution of water. This role embodies principles of fairness and equity, ensuring all community members receive their rightful share based on time-honoured schedules and agreements. The Mirab's position is culturally significant, reinforcing the importance of communal responsibility and sustainability [15].

Communal maintenance of qanats has historically bonded communities in Yazd. Regular maintenance activities, such as the annual cleaning of qanat tunnels, involve entire villages and are often followed by communal meals and celebrations. These rituals underscore the collective responsibility towards sustainable water use and reinforce social cohesion. This system highlights the ethical and cultural dimensions of water management in Yazd, blending Iranian, Islamic, and modern values [7].

A notable instance of community-based involvement is the Takab Rural District project in Kerman, spearheaded by Nina Aminzadeh. This endeavour aimed to foster sustainable livelihoods by rejuvenating qanats through participatory water resources management. Supported by the Centre for Sustainable Development and Environment (CENESTA), a Livelihoods Fund was established, engaging women within the community. Through the women's cooperative, Gojino, a portion of profits is earmarked for qanat revitalization, showcasing a successful fusion of community development and resource conservation [11].

3 Qanats: A Sustainable Solution

Qanats' innovative design ensures efficient water management, passive cooling, and biodiversity support, making them environmentally beneficial and resilient.

3.1 Ecological Benefits and Environmental Impact

Yazd's qanats are remarkable for transporting water through underground tunnels and significantly reducing evaporation. This method maintains stable groundwater levels, preventing depletion and ensuring a consistent water supply for the region.

Also, the qanat system promotes environmental stability by preventing land degradation and desertification. By safeguarding water from pollutants and contaminants, qanats help preserve the quality of both water and soil.

Qanats also play a crucial role in managing excess water during heavy rainfall, reducing flood risks, and preventing waterlogging. They help balance soil salinity by providing a steady flow of fresh water, which dilutes salt concentrations in the soil. This function is especially important in arid regions where high evaporation rates can lead to salt build up, adversely affecting crop growth.

3.2 *Passive Cooling and Humidification in Urban Areas*

In urban areas like Yazd, qanats play a crucial role in natural cooling and humidification. These ancient underground aqueducts, originating over 3,000 years ago in Persia, utilize the cooler temperatures below the surface to channel cool water into the city. As the water flows through the qanats, it cools the surrounding earth and air, creating a more comfortable urban environment [1]. This system is especially effective in arid regions, where water scarcity and extreme heat are common (see Fig. 1).

Windcatchers, or badgirs, are traditional ventilation structures often integrated with qanats. These architectural elements capture hot air at the top and channel it down through a canal, cooling it as it passes over water pools in basements. This

Fig. 1 The Shesh Badgir Ab Anbar in Yazd features six windcatchers for natural ventilation and cooling (Photo by Author)

process delivers cooler air into buildings, significantly reducing the need for modern air conditioning. The combination of qanats and windcatchers creates an efficient passive cooling system, which reduces energy consumption and lowers greenhouse gas emissions. This method stabilizes indoor temperatures, keeping them within a comfortable range, and enhances the overall air quality by maintaining higher humidity levels.

The Shesh Badgir Ab Anbar in Yazd is an exemplary model of traditional Persian architecture integrated with the qanat system. Located on Shahid Beheshti Street, this 180-year-old Qajar-era cistern features six windcatchers, which play a crucial role in cooling the air and maintaining the water temperature. Serving as an entrance to the qanat, it includes stairways leading to underground water reservoirs. The vaulted brickwork entrance and dual stairways highlight both the functionality and aesthetic appeal of the structure. Additionally, the adjoining Hosseiniyeh serves as a communal gathering space, emphasizing the social and cultural significance of these structures in promoting community cohesion and resilience.

3.3 Contribution to Biodiversity and Green Spaces

In Yazd, qanats foster a rich array of plant and animal life. These systems sustain diverse vegetation, including date palms, pomegranates, and various herbs, which in turn provide food and shelter for birds, insects, and other wildlife. The resulting biodiverse areas contribute to ecological balance, fostering more resilient and sustainable ecosystems. Moayedfar and Fatemi [17] note that the presence of aquatic species such as fish, crabs, and shrimp in qanats also enhances their appeal to tourists.

Historical gardens like the Dowlatabad Garden exemplify the success of qanats in creating lush green spaces in desert environments. These gardens serve as cultural and historical landmarks, preserving local biodiversity while showcasing the efficacy of qanats in supporting agriculture. Moreover, qanats contribute significantly to food security in the region by enabling the cultivation of various crops (see Fig. 2).

The presence of garden pits, such as Gudalbaghche, in Yazd, suggests access to underground water aligned with the water level of qanats. This feature, common in central and desert cities in Iran, helps regulate heat and creates a cooler atmosphere within homes [19].

Beyond their ecological and agricultural benefits, qanats also promote social cohesion by providing communal spaces centred around water features, encouraging community interactions. Integrating qanats into urban infrastructure not only supports sustainable water management but also promotes environmental conservation and enhances overall quality of life. This illustrates how traditional engineering approaches can effectively address modern urban challenges.

Fig. 2 Dowlatabad Garden in Yazd, sustained by qanats, exemplifying lush green spaces and biodiversity in a desert environment (Photo by Author)

4 Integration into Modern Urban Infrastructure

Integration of qanats into contemporary urban landscapes presents challenges that need to be addressed, including the conservation of these ancient structures and the adoption of technological advancements to enhance water management strategies and environmental resilience in cities.

4.1 Addressing Challenges and Degradation

The qanat system, historically crucial for water management in Yazd, is now facing significant challenges and degradation. Prolonged droughts have severely impacted groundwater resources, leading to the drying up of approximately 800 qanats, which accounts for 25% of the total in Yazd. Compounding this issue is the fact that 83% of Yazd's qanats are situated along valleys or mountainous areas, rendering them vulnerable to fluctuations in rainfall and unable to access deeper water-bearing zones [22].

Illegal and excessive water withdrawal, facilitated by energy-consuming pumps, further exacerbates the depletion of these resources. This has resulted in a notable increase in wells, posing a threat to the sustainability of qanats. Urban expansion and neglect also pose risks to many traditional qanats, underscoring the need for integrated conservation strategies. While modelling the earth fissure hazard is essential for identifying vulnerable groundwater areas and ensuring informed water management [4], it is important to recognize that qanats themselves are not direct hazards. They are significant cultural heritage sites, and attributing hazards directly to them is unwarranted [21].

Technically, the dewatering of qanats due to the overuse of deep wells presents a considerable challenge. Modern alternatives like deep wells, which pump groundwater faster than its natural replenishment rate, are favoured for their speed but lack the sustainability of traditional qanats. Regular maintenance, erosion control, and protection from contamination are crucial for preserving the structural integrity and functionality of qanats. Additionally, establishing a fair water allocation system is essential to meet the needs of all users and ensure community support for the maintenance of qanats.

4.2 Preservation Challenges and Solutions

Preserving the qanats in Yazd is crucial yet challenging. The primary issues stem from urban migration and the decreasing number of skilled workers. Regular maintenance, like tunnel cleaning and sediment removal, is essential to prevent these ancient systems from deteriorating. Community-based programs can play a vital role in engaging residents in maintenance, ensuring both sustainability and the transfer of knowledge. Local councils often organize cleaning and repair sessions, which effectively maintain these systems.

Another major challenge is the timely detection of leaks and blockages. Innovative solutions, such as the Smart Qanat Initiative, use sensors and remote monitoring to provide real-time data. Maintaining the structural integrity of qanats also requires using traditional materials and techniques. Extensive restoration using these traditional methods has ensured their continued operation.

In some regions, international collaboration has improved water management and preservation through shared best practices. Stricter regulations on urban planning and waste management, along with public awareness campaigns, can protect these vital systems. Combining traditional practices with modern innovations is essential for preserving Yazd's qanats for future generations.

4.3 Technological Innovations in Water Management

The traditional techniques used to build and maintain qanats are time-consuming, making modern alternatives like deep wells more appealing.

Technological innovations offer promising solutions. The use of nanotechnology in water and wastewater treatment improves treatment efficiency and enables the use of unconventional water sources. Nanomaterials, such as Nano adsorbents and nanomembranes, enhance water purification processes, although commercialization barriers and regulatory challenges remain [8]. Additionally, smart water management technologies leveraging the Internet of Things (IoT) enable real-time monitoring and control of water usage, promoting more efficient water management and contributing to urban sustainability [20].

Additionally, regulatory frameworks and public health concerns also slow down the implementation of new technologies. Experiences from countries like the Netherlands demonstrate that stakeholder support, economic incentives, and systematic development of water-related services are essential for promoting new technologies [12].

Governance is transforming, moving towards public management with a focus on societal engagement. This approach acknowledges the intricate connections between water resources and society, fostering innovations that take these dynamics into account. Legal frameworks and policies supporting water conservation and equitable allocation play a pivotal role in driving technological advancements in water management [14]. A comprehensive approach encompassing technological, regulatory, and societal factors is indispensable for tackling the challenges and safeguarding the qanats of Yazd.

5 Policy Frameworks and Future Directions

Effective policy and governance are crucial for the preservation and sustainability of qanats, particularly in regions like Yazd. Legal protections play a vital role, with specific laws enacted to prevent the destruction and misuse of these ancient water systems. Historically, a body of custom and law (shari'a) was developed to regulate qanat use, ensuring fair water distribution and protecting the rights of qanat owners. For example, the law of harim ("borders") prohibited the sinking of new mother wells within one kilometre of existing qanats [6].

Contemporary preservation efforts encompass integrating qanat conservation into diverse development plans, spanning agriculture, urban development, industry, and natural resource projects. This holistic approach ensures the consideration and safeguarding of qanats and their surrounding landscapes [11]. Legal protection for qanats is provided by the Law for the Protection of National Heritage (1930), which includes them on the National Register, alongside national laws and regulations safeguarding underground water resources and catchment areas.

Organizations such as the International Centre for Qanats and Historic Hydraulic Structures, operating under UNESCO's auspices, undertake detailed mapping and documentation to preserve maintenance techniques and historical knowledge. This data, held by local branches of ICHHTO, is made accessible to community members, fostering a sense of ownership and responsibility among residents. Traditional communal management systems, supported by government authorities, play a vital role in the ongoing maintenance and protection of these systems [10].

Through a combination of legal protections, integrated development planning, and community involvement, the preservation and sustainable management of qanats can be effectively achieved, ensuring that these crucial water systems continue to support both people and the environment.

Furthermore, incorporating qanats into contemporary urban planning can provide a reliable water source for urban populations while preserving cultural heritage. Modern technology can enhance the efficiency and sustainability of these ancient systems.

Urban development plans must include protective measures to ensure qanats are not disrupted by construction activities. Zoning policies should safeguard qanat corridors and adjacent agricultural lands, promoting sustainable land use practices. Integrating qanats with modern water distribution infrastructure, including supplementary reservoirs and channels, can significantly improve water storage and distribution efficiency.

Utilizing qanat water for creating and maintaining urban green spaces, parks, and gardens can contribute to urban cooling and improve air quality. Additionally, this integrated management can reduce energy consumption for cooling by mitigating the urban heat island effect.

Community engagement stands as a cornerstone for safeguarding and upkeeping qanat systems. By implementing educational programs and incentives, a sense of ownership and responsibility can flourish among residents. Furthermore, promoting the revitalization of qanats holds the potential to stimulate tourism and bolster agriculture. Initiating eco-tourism endeavours centred around qanat sites not only draws visitors and generates income but also ensures a reliable water source for agricultural lands, thereby supporting local food production. This holistic approach guarantees the enduring support of qanats for both agricultural and urban sustainability, preserving their functionality and averting collapse.

6 Conclusion

The qanat system stands as an indispensable pillar of Yazd's climate resilience, embodying an ancient yet remarkably effective approach to water management. These subterranean channels, driven by gravity, not only minimize evaporation but also uphold water quality, offering a sustainable solution amidst the arid landscapes of Yazd. Their enduring legacy is underscored by the ability to deliver a consistent and dependable water supply, confronting the perpetual challenge of water scarcity head-on.

Crucially, qanats are vital for sustaining Yazd's ecological and economic equilibrium. Beyond supporting agriculture, a linchpin of local food security and livelihoods, they serve as bulwarks against the escalating impacts of climate-induced droughts. To safeguard and optimize the utility of qanats, the integration of modern technologies with traditional practices is imperative. By incorporating sensor-based systems for real-time monitoring and involving local communities in maintenance programs, we not only ensure the long-term functionality of qanats but also foster a sense of shared responsibility and stewardship among residents. Looking forward, the timeless principles underpinning the qanat system offer a roadmap for developing hybrid water management models that blend historical wisdom with modern innovation. This holistic approach not only addresses contemporary water scarcity challenges but also preserves the rich cultural heritage of regions like Yazd, providing a sustainable path forward for future generations.

References

1. Abouei R (2006) Conservation of Badgirs and Qanats in Yazd, Central Iran. https://www.academia.edu/download/106342021/PLEA2006_PAPER628.pdf
2. Azizi-Jalilian M, Shayesteh K, Danehkar A, Salmanmahiny A (2020) A new ecosystem-based land classification of Iran for conservation goals. Environ Monit Assess 192:182. https://doi.org/10.1007/s10661-020-8145-1
3. Boustani F (2008) Sustainable water utilization in arid region of Iran by Qanats. Int J Civ Environ Eng 2(7):152–155
4. Choubin B, Mosavi A, Alamdarloo EH, Hosseini FS, Shamshirband S, Dashtekian K, Ghamisi P (2019) Earth fissure hazard prediction using machine learning models. Environ Res 179(Pt A):108770. https://doi.org/10.1016/j.envres.2019.108770
5. Dehghani-Sanij A, Bahadori MN (2021) The history of traditional Iranian ice-houses. In Ice-Houses, pp 131–151
6. English PW (1968) The origin and spread of Qanats in the old world. Proc Am Philos Soc 112(3):170–181
7. Foltz R (2002) Iran's water crisis: cultural, political, and ethical dimensions. J Agric Environ Ethics 15:357–380. https://doi.org/10.1023/A:1021268621490
8. Gehrke I, Geiser A, Somborn-Schulz A (2015) Innovations in nanotechnology for water treatment. Nanotechnol Sci Appl 8:1–17. https://doi.org/10.2147/NSA.S43773
9. Hamidian A, Ghorbani M, Abdolshahnejad M, Abdolshahnejad A (2015) Retracted: Qanat, traditional eco-technology for irrigation and water management. Agric Agric Sci Procedia 4:119–125. https://doi.org/10.1016/j.aaspro.2015.03.014
10. ICOMOS (2016) The Persian Qanat (Islamic Republic of Iran) No. 1506: evaluation report. https://whc.unesco.org/en/list/1506/documents/
11. Karimian A, Zivdar N, Ricard B (2021) Participatory conservation of Persian Qanats. In: Persian paradises at peril, pp 185–201. Miranda AD (2014) Water-harvesting tradition in Syrian steppe. Procedia Eng 89:1522–1528. https://doi.org/10.1016/j.proeng.2014.11.446
12. Krozer Y, Hophmayer-Tokich S, Meerendonk H, Tijsma S, Vos E (2010) Innovations in the water chain—experiences in the Netherlands. J Clean Prod 18:439–446. https://doi.org/10.1016/J.JCLEPRO.2009.11.013
13. Labbaf Khaneiki M, Al-Ghafri AS, Klöve B, Haghighi AT (2022) Sustainability and virtual water: the lessons of history. Geogr Sustain 3:358–365. https://doi.org/10.1016/j.geosus.2022.11.005
14. Larson R (2020) Just add water: Solving the world's problems using its most precious resource. Oxford University Press
15. Megdiche-Kharrat F, Ragala R, Moussa M (2019) Promoting a sustainable traditional technique of aquifer water acquisition common to arid lands: a case study of Ghassem Abad Qanat in Yazd Province. Water Supply 19:527–535. https://doi.org/10.2166/WS.2018.098
16. Mehrjardi R, Mahmoodi S, Heidari A, Akbarzadeh A (2009) Micromorphological evidences of climatic change in Yazd region, Iran. J Mt Sci 6:162–172. https://doi.org/10.1007/S11629-009-1033-3
17. Moayedfar S, Fatemi M (2021) Sustainable tourism development in historic cities of arid regions with the revival of Qanat (Yazd City). GeoJ Tour Geosites 35(2):428–436. https://doi.org/10.30892/gtg.35222-669

18. Motiee H, McBean E, Semsar A, Gharabaghi B, Ghomashchi V (2007) Assessment of the contributions of traditional Qanats in sustainable water resources management. Int J Water Resour Dev 22(4):575–588. https://doi.org/10.1080/07900620600551304
19. Omidvari A, Zivdar M (2021) An analysis on the morphological typology of Godal Baghche courtyards in the central plateau of Iran; case study: traditional houses of Yazd. Stud Islam Iran City 11(44):189–210
20. Palermo S, Maiolo M, Brusco A, Turco M, Pirouz B, Greco E, Spezzano G, Piro P (2022) Smart technologies for water resource management: an overview. Sensors (Basel, Switzerland) 22. https://doi.org/10.3390/s22166225
21. Parise M (2016) Qanat is not a hazard. Environ Earth Sci 75:1–5. https://doi.org/10.1007/s12665-016-6284-z
22. UNESCO (2007) Groundwater for emergency situations. United Nations Educational, Scientific and Cultural Organization. Paris, France
23. Zarei M, Ashkezari S, Yari M (2017) The investigation of the function of the central courtyard in moderating the harsh environmental conditions of a hot and dry climate (Case study: City of Yazd, Iran). Spatium (38):1–9. https://doi.org/10.2298/spat1738001z

Resilience in Crisis: The Underground as a Refuge

Unveiling the Subterranean Frontier: Leveraging Underground Spaces for Urban Resilience and Expansion

Michael Rehbock and Antonia Cornaro

Abstract This chapter highlights the crucial role of subterranean environments in fortifying cities against climate change, particularly flooding and extreme weather. Underground spaces house essential services like power distribution, water supply, and data centres, enhancing urban resilience during crises. Large underground caverns offer new opportunities for logistics, production, utilities, storage, and entertainment. Strategically deploying these spaces can significantly increase urban capacity and free up surface areas for diverse uses. However, building subterranean infrastructure presents challenges, including structural integrity, construction methodologies, logistical complexities, and geotechnical issues. Ensuring fire safety and integrating underground developments with existing urban areas require careful planning and execution. This chapter examines the growing trend of underground utilisation, outlining the opportunities, challenges, and essential considerations for leveraging subterranean spaces to boost urban resilience and expansion. It highlights exemplary long-term cavern development strategies, such as those in Hong Kong. Hong Kong's approach involves meticulous geological assessments, innovative construction techniques, and stringent safety protocols. By leveraging its unique topography and geology, Hong Kong has transformed its underground spaces into thriving hubs for commerce, recreation, and infrastructure. This case study serves as a model for cities worldwide, demonstrating the transformative potential of strategic underground development in fostering sustainable urban growth and resilience.

Keywords Underground · Subterreanean · Cavern · Resilience · Modular expansion

---

M. Rehbock
Amberg Engineering AG, Expert, Trockenloostrasse 15, 8105 Regensdorf-Watt, Switzerland
e-mail: mrehbock@amberg.ch

A. Cornaro (✉)
Urban Planner, Expert and Lecturer Underground Space, Amberg Engineering AG, Trockenloostrasse 15, 8105 Regensdorf-Watt, Switzerland
e-mail: acornaro@amberg.ch

© The Author(s), under exclusive license to Springer Nature Singapore Pte Ltd. 2025
M. Agrawal (ed.), *Underground Spaces for Climate Resilience and Sustainability*, Advances in 21st Century Human Settlements,
https://doi.org/10.1007/978-981-96-0547-7_9

1 Introduction

Urban areas worldwide are facing unprecedented challenges due to rapid population growth and increasing land demand. As cities expand, surface space becomes scarce and precious. Moreover, climate change, with its associated risks of flooding and extreme weather events, underscores the need for enhanced urban resilience. Utilising underground spaces has emerged as a compelling solution to these issues. By tapping into subterranean environments, cities can manage spatial constraints and improve their disaster recovery capabilities.

Urban underground space development encompasses a diverse range of applications, from essential infrastructure like power distribution and water supply to commercial ventures such as data centres and entertainment facilities. Large-span underground caverns offer unique opportunities for urban expansion and land use optimisation. However, realising the full potential of underground spaces requires strategic planning, innovative engineering solutions, and a holistic approach to urban development.

Underground spaces enhance urban resilience by hosting essential services like power distribution, water supply, and data centres, ensuring these functions remain operational during crises. Additionally, large underground caverns can be utilised for logistics, production, utilities, storage, and entertainment, providing multifaceted applications.

As climate change intensifies, the need for cities to protect themselves from extreme weather events increases. Underground spaces can help cities become more resilient by hosting vital services. Constructing large caverns in urban areas, where geological conditions permit, frees up surface space and provides strategic additional capacity for diverse uses.

However, realising subterranean infrastructure poses significant challenges, including structural integrity, construction methodologies, logistical complexities, and geotechnical considerations. Additionally, ensuring fire safety and seamlessly integrating these underground developments with existing urban infrastructure requires meticulous planning and execution. Specific challenges encompass structural issues, advanced construction technologies, logistics, geotechnical obstacles, fire safety, and integration with current urban systems.

2 Urban Underground Space Development: A Global Snapshot

Cities are expanding vertically as well as into the sea, leading to a surge in the quest for urban prime space. Some cities have embarked on master planning of their underground space which addresses these challenges. Large-span underground caverns represent a transformative solution to the spatial constraints faced by urban areas. Whether repurposed from existing underground quarries or purposefully excavated, these caverns offer significant potential for urban expansion and optimisation of land use.

These underground caverns offer diverse applications including logistics, production, utilities, storage, and entertainment. In Singapore, Hong Kong, and Switzerland, several caverns have been developed for underground fuel storage, quarries, laboratories, factories, sewage treatment plants, columbaries, modern logistics, and data centres. However, cavern development requires careful consideration of geological features, structural integrity, and logistical planning.

Advancements in engineering technologies and construction methodologies have enabled the design and construction of large-span caverns tailored to specific urban needs. By maximising the utilisation of rock mass volume and minimising geomechanical interactions between caverns, engineers can optimise the layout of underground spaces. The integration of modern characterisation and analysis tools enables the identification and mitigation of potential risks associated with cavern development.

Hong Kong, Singapore, Helsinki, Oslo, and other cities across the world have initiated or are well ahead in the game of developing and implementing ambitious plans for their subspace.

This chapter focuses on Hong Kong's comprehensive approach to subterranean utilisation via long-span caverns leveraging its unique topographical and geological features, successfully transforming its underground spaces into thriving hubs for commerce, recreation, and infrastructure. It also briefly highlights the underground space planning approach adopted by Singapore. These two examples set a precedent for cities worldwide, illustrating the transformative potential of strategic underground development in fostering sustainable urban growth and resilience.

2.1 Cavern Development in Hong Kong

Land supply is a pressing issue that constrains the sustainable development of Hong Kong. In the 1990s, Hong Kong initiated a cavern development planning and development program that led to a public outreach program and a subsequent cavern development master plan [2]. It aims to relocate major public facilities underground, thereby releasing valuable land above ground for alternative uses.

Since 2010, the Civil Engineering and Development Department (CEDD) has supported the government in establishing technical policies to facilitate cavern development as a measure to increase land supply. Key initiatives include:

- Realisation of new cavern developments for concrete batching, rock processing, and asphalt production plants.
- Relocation of the Public Works Central Laboratory to caverns.
- Pilot Planning and Engineering Study on the development of selected strategic cavern areas with priorities for data centres, modern logistics facilities, and columbariums.

The cavern development plan outlines 48 major areas for strategic cavern development, ranging from 30 to 200 hectare (ha), as depicted in Fig. 1. It entails relocating significant public facilities into large-span caverns. The first facility, the Sha Tin Sewage treatment plant will be placed underground and therefore release 28 ha of land. Additionally, feasibility and pilot studies are being led to examine the potential relocation of laboratories, data centres, columbariums, and other innovative ventures into caverns. Both studies are currently being conducted in collaboration with Amberg Engineering (see Fig. 2).

Hong Kong has implemented a comprehensive approach to subterranean utilisation, including meticulous geological assessments, innovative construction techniques, and stringent safety protocols. The city has successfully transformed its underground spaces into thriving hubs for commerce, recreation, and infrastructure.

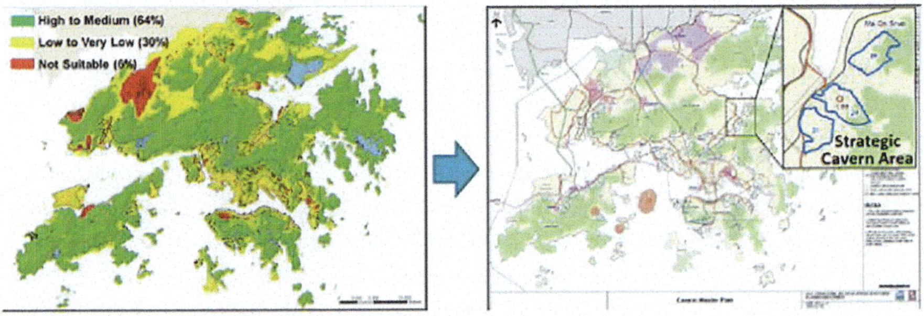

Fig. 1 Strategic cavern areas in Hong Kong, CEED, GEO [2]

Fig. 2 Relocation of Sha Tin sewage treatment plant into a cavern, Hong Kong, CEED, GEO

2.2 Singapore's Strategic Approach to Underground Space Planning and Management

Singapore has embarked on a comprehensive underground master plan to optimise urban space, alleviate surface congestion, and enhance the living environment by relocating infrastructure and industrial uses to underground rock caverns. This strategic initiative aims to address land scarcity and improve urban resilience against climate change, particularly by mitigating flooding and extreme weather impacts.

Singapore's approach to underground space utilisation includes learning from global best practices. In 2014, the Urban Redevelopment Authority (URA) commissioned a global benchmarking study to explore international standards in underground space planning and management—across Beijing, Helsinki, Hong Kong, London, Montréal, Paris, Seoul, Shanghai, Taipei, and Tokyo.

One key recommendation was to create and maintain a central data bank of complete, accurate, and up-to-date information on geology and existing and proposed underground developments. This would reduce uncertainty in subterranean planning, management, and construction. Strengthening public–private partnerships, as evident in Tokyo and Montréal, was also recommended to encourage the built environment industry to undertake underground projects despite higher costs and risks.

Findings from the study have laid the groundwork for developing an underground master plan for detailed district planning, guiding large-scale and plot-specific developments. The URA aims to release an underground master plan for pilot areas, enhancing planning and management strategies for underground development. These strategies offer exciting opportunities to improve the quality of Singapore's future living environment, demonstrating the transformative potential of strategic underground development in fostering sustainable urban growth and resilience.

In March 2019, the Urban Redevelopment Authority (URA) of Singapore unveiled the Draft Master Plan 2019, emphasising the strategic use of underground space to meet the city's growing needs. Singapore's extensive underground networks for utility infrastructure highlight the importance of reliable subsurface information. The Digital Underground project, a collaboration between the Singapore Land Authority and the Singapore-ETH Centre, aims to create a digital twin of all subsurface utilities.

Key projects in the feasibility stage, such as the Underground Science City, the relocation of utility plants to underground rock caverns, and the Gali Batu underground goods movement system, are expected to advance with new legal policies focused on placing such facilities underground.

By embracing innovative engineering solutions, strategic planning, and a holistic approach to urban development, Singapore is setting a global example in maximising the use of its underground space. This approach not only addresses the city-state's land scarcity but also ensures a more resilient and sustainable urban future.

3 Cavern Development and Management

This section spotlights real-world examples of urban cavern design and development from Europe and Southeast Asia. From industrial quarries to data centres and underground farming facilities, these projects demonstrate a diverse range of applications for underground spaces.

In Switzerland, for instance, underground quarries designed and supported by engineering firms like Amberg Engineering have enabled the extraction of rock materials for various industrial purposes. Similarly, Singapore's efforts to relocate public utilities underground demonstrate the potential for optimising space utilisation and enhancing energy efficiency in densely populated urban environments. These examples highlight the importance of collaboration between government agencies, private sector stakeholders, and engineering firms in realising the full potential of urban underground spaces.

Underground mining, while contingent on favourable geological and environmental conditions, presents significant economic advantages and unique benefits for urban areas despite its higher initial costs and safety challenges. The key advantages of operating an underground quarry in urban settings include:

- Underground quarries reduce surface disruption, thereby addressing zoning issues and minimising environmental impacts.
- Unlike surface quarries that require extensive rehabilitation (e.g. the 18 years of restoration at Shek O Quarry in Hong Kong), underground quarries avoid these lengthy and costly processes.
- Underground quarries can tap into reserves beneath the quarry floor, under pit slopes, or on adjoining properties, thereby maximising resource extraction.
- Underground operations benefit from a stable climate, free from external weather conditions such as rainfall, frost, or strong winds.

To successfully operate an underground quarry, several prime requirements must be met [1]:

- The underground mining must not affect the morphology of the surface in terms of lowering the ground surface. The overall stability of the system must be ensured.
- High profitability and maximum efficiency are achieved by high mechanisation and recovery of rock materials.
- A maximum possible amount of rock material must be exploited.
- Strategic future use of the cavern and tunnel systems.
- Minimised environmental impact on the surrounding areas.

3.1 Urban Caverns and Quarries of Switzerland

In Switzerland, three underground quarries are operational. Designed and supported by Amberg Engineering, these quarries demonstrate the successful implementation of these principles and requirements:

- The Schollberg, extracts 100,000 m^3 of limestone in a year, using a room and pillar layout and cavern cross-section of 216 m^2 (see Fig. 5)
- The Zingel mine extracts 150,000 m^3 of silicate limestone per year, with caverns 12 m wide and 25 m high, resulting in a cross-section area of 300 m^2.
- The Läntigen quarry produces approximately 100,000 m^3 of limestone per year, with caverns aligned in a complex three-dimensional layout, featuring a width of 15 m and a height of 18 m, resulting in a cross-section area of 324 m^2.

These three production-oriented quarries in Switzerland operate without excavation support, following a cost-effective design methodology. This approach ensures reliable, stable and efficient operations by adhering to the following three steps:

- Sound ground characterization: Identify major input variables beforehand, including intact rock strength, degree of fracturing, groundwater presence, permeability, and discontinuity orientation. This step ensures a thorough understanding of the geological conditions, critical for safe and efficient excavation.
- Determine pillar width: to ensure stable excavations. This decision is based on commonly used pillar strength and pillar loading theories, balancing stability and economic feasibility.
- Conduct a parametric study using Unwedge to determine the optimal orientation of the main caverns. The goal is to minimise the volume of unstable blocks while enhancing overall stability and safety.

These steps, grounded in sound engineering principles and supported by robust geological and stability analyses, enable the effective and safe operation of underground quarries. By prioritising ground characterisation, strategic pillar width selection, and parametric stability studies, these quarries maintain low-cost production while ensuring excavation stability and safety.

Furthermore, short-term stability is ensured by good workmanship (defined as quality blasting and systematic extrados scaling after blasting), while long-term stability is an inherent aspect of the design and is achieved solely by an optimal layout of the excavation. Only in cases where the conditions depart from the anticipated baseline in a significant manner (presence of faults, stress concentrations observable by spalling), the excavation is supported by shotcrete and spot bolting.

3.1.1 Waferfab Facility

The Waferfab manufacturing facility is situated on the western valley flank of the St. Gallen Rhine valley at the foot of the Gonzen. The decision to construct the

Fig. 3 Aerial view of "Waferfab" area (left), caverns (right), Amberg Engineering

facility underground was based on the economic advantage of decreased installation costs associated with vibration-sensitive equipment required for chip production (see Fig. 3).

The facility includes surface office buildings as well as underground manufacturing caverns, connected by two service tunnels on different access levels of the cavern.

Due to safety and transport requirements, it was necessary to situate the structures near the surface, resulting in the overburden ranging from 20 to 70 m above the main cavern diminishing to 7 m at the end of the "Litho cavern". The dimensions of the main production cavern are 100 m × 18 m × 18 m (l-s-h) while the "Litho cavern" (56 m × 18 m × 10 m) is situated perpendicular to the main production cavern.

Geological investigation as well as the experience with projects in the same rock mass (Quinten Limestone) have shown that gravitation and stress-induced block failure at higher overburdens/stresses are common. These are accompanied by wall buckling in case of adverse bedding orientation. In addition, spalling phenomena have been documented as well—at the bifurcations of the chambers of the adjacent, abandoned mine.

Based on this knowledge and additional geotechnical analysis, a uniform permanent support consisting of glass fibre 8–10 m long reinforced rock bolts and 20 cm of fibre-reinforced shotcrete has been used. The occurrences of unstable blocks or other adverse aspects of unstable behaviour have been coped with by reducing the round length and installing the support as soon as possible.

Major open discontinuities have been grouted to generate a tight load-bearing "ring" around the tunnel and prevent stress concentrations. Due to the groundwater level approximately 2–3 m below invert and directly connected with the Rhine River level, the entire facility has been designed and constructed with a permanent single-shell shotcrete lining. The owner and operator of the Waferfab facility chose an underground solution for a vibration-free environment, where he can operate a clean room in a completely stable environment (see Fig. 4).

Fig. 4 The Schollberg quarry, Amberg Engineering

3.1.2 Hagerbach Test Gallery

In Switzerland, two recent innovative prototypes have been developed to showcase the potential of underground spaces under real conditions:

- Underground data centre: Housed in a cavern of the Hagerbach Test Gallery, this data centre leverages the stable and cool underground climate for enhanced energy efficiency and operational cost savings.
- Underground farming: An aquaponics system that combines trout fish farming with hydroponic plant cultivation, demonstrating sustainable agricultural practices in an underground environment.

These projects benefit from the engineering expertise of Amberg Engineering, which collaborates with SCAUT, the Swiss Center of Applied Underground Technologies, at Hagerbach. The climatic conditions within caverns provide significant energy efficiency and economic benefits, making them ideal for such innovative applications.

Hagerbach Test Gallery is an underground infrastructure, spanning over 5.5 kms and featuring diverse rock formations and installations, offering optimal conditions for research, testing, and practical demonstrations. Amberg has been responsible for engineering services at this test facility since its foundation in 1970. This includes structural engineering support, advanced calculations, hydrogeological assessments, and continuous technical documentation.

A three-dimensional BIM model of the facility is currently being developed to optimise future planning and operational use of the tunnel system (see Fig. 5).

Fig. 5 Prototype for edge computing data centre in cavern in Hagerbach, Switzerland, image: SCAUT

4 Fire Safety in Underground Structures

The safety of underground structures is paramount and must be meticulously considered during all phases of planning, design, and construction. Fire safety, in particular, presents unique challenges, such as maintaining structural integrity, selecting suitable construction methods, and addressing the distinct characteristics of subterranean environments. While existing building codes offer a baseline for fire safety, underground structures necessitate specialised approaches to effectively mitigate their specific risks.

A performance-based fire engineering design provides a comprehensive strategy for addressing fire-related threats and ensuring safety in underground settings. This approach includes:

- Fire-resistant construction materials: Utilising materials that can withstand high temperatures and prevent the spread of fire.
- Effective smoke control strategies: Implementing systems to manage and control smoke movement, ensuring visibility and air quality.
- Adequate means of egress: Designing multiple, accessible escape routes to facilitate safe evacuation in case of an emergency.

These measures help minimise the potential impact of fires in underground structures. Additionally, continuous monitoring and maintenance programs are vital to ensuring the long-term safety and resilience of these environments. By adopting a tailored, performance-based approach to fire safety, engineers can effectively protect underground structures from fire-related hazards.

5 Conclusion

This chapter has delved into the expanding field of underground space utilisation, highlighting its potential to address urban challenges and promote resilient, sustainable growth. By leveraging subterranean environments, cities can overcome spatial constraints, optimise land use, and bolster their ability to withstand and recover from disasters. However, unlocking the full potential of these spaces necessitates strategic planning, innovative engineering, and a steadfast commitment to safety and sustainability. Embracing underground development as a vital component of urban planning and design will enable cities to create vibrant, resilient, and sustainable environments for future generations.

6 Areas of Further Exploration

As urbanisation and environmental challenges persist, the strategic development of underground spaces will be pivotal in shaping sustainable urban growth and resilience [3]. Future initiatives should prioritise collaborative planning, cutting-edge technologies, and robust policy frameworks to fully exploit the potential of subterranean spaces. This comprehensive approach underscores the importance of underground space utilisation in tackling urban challenges and fostering resilient, sustainable cities.

References

1. Amberg F (10–12 Dec 2014) Quarries and caverns, underground space for industrial and R&D uses. In: 1st Asia-Pacific conference on complex systems design & management. National University of Singapore, pp 1–47
2. Cavern Master Plan of Hong Kong, by Edward LO, PlanD & Tony HO, CEDD, ITA Tunnelling Awards, 2015
3. ITACUS, Admiraal H, Cornaro A (2016) Why underground space should be included in urban planning policy—and how this will enhance an urban underground future. Tunnelling and Underground Space Technology

Underground and Beyond: The Road to Resilience

Patricia Lamas Alvarez

Abstract As urbanization accelerates, cities are increasingly challenged to provide adequate infrastructure while addressing environmental impacts and space constraints. This chapter explores the potential of underground infrastructure as a solution to these challenges. By leveraging tunnelling and advanced construction techniques, cities can create new spaces in saturated urban areas, promote sustainable practices, and enhance the resilience of urban environments. This shift towards subterranean development requires a transformation in both social mindset and regulatory frameworks. The chapter also delves into the innovative use of real-time monitoring systems, artificial intelligence, and predictive analytics to optimize maintenance and pre-emptively address issues, heralding a new era in construction. Through an examination of underground infrastructure's benefits and the necessary policy changes, this chapter outlines a path towards more sustainable and resilient urban development.

1 Introduction

The development of new infrastructure in today's global context faces several challenges, particularly when ensuring resilience. The modern urban environment is characterized by a heavily constrained network, presenting many restrictions for transportation development, such as space limitations, the conglomeration of existing structures, surface traffic, and disturbances to the everyday life of city residents. Underground infrastructure emerges as an ideal alternative solution for addressing these issues.

This chapter aims to explore possibilities within the underground space. Underground infrastructure provides various opportunities for sustainable practices, such as utilizing self-sustaining energy sources through geothermal activation and employing innovative construction methods to reduce the carbon footprint. Additionally, it

P. L. Alvarez (✉)
Amberg Engineering, Brussels, Belgium
e-mail: plamas@ambergengineering.com

© The Author(s), under exclusive license to Springer Nature Singapore Pte Ltd. 2025
M. Agrawal (ed.), *Underground Spaces for Climate Resilience and Sustainability*,
Advances in 21st Century Human Settlements,
https://doi.org/10.1007/978-981-96-0547-7_10

allows for efficient use of surface space, promoting social cohesion and civic engagement.

The following sections will discuss the necessary transformations in regulatory and policy contexts required to transition to this new paradigm of infrastructure development, imagining a future where underground solutions play a pivotal role in urban planning.

2 Underground, a New Landscape for Infrastructure Development

The private engineering sector is gradually embracing the challenge of enhancing infrastructure resiliency. Environmental considerations have become central to tunnelling projects, with the primary challenge being the timely identification of hazards that could compromise project durability.

Once the hazards are identified, expert collaboration and rigorous contextual analysis enable the industry to shift towards more sustainable development patterns. Adopting a circular economy approach, which emphasizes the reuse, recycling, and repurposing of materials, is crucial in reducing the environmental footprint of projects.

Embracing a circular economy approach promotes the reuse, recycling, and repurposing of materials to reduce the overall environmental footprint of the project. Exploring the possibilities of underground space presents a promising path to ensuring the livability of urban environments. By integrating these practices, the engineering sector can lead the way in developing resilient, sustainable infrastructure that meets the demands of modern urbanization.

2.1 New Initiatives and Criteria Assessment

Improving energy efficiency is crucial for building resilient infrastructure. Achieving this goal requires a supportive policy framework that includes:

- Fair Green Certifications and Transparent Labelling: Establishing standardized certifications and clear labelling to guide sustainable practices.
- Incentivizing Data-Driven Design Tools: Promoting the use of advanced design tools that leverage data for optimizing energy efficiency.
- Promoting Investments in Sustainable Materials: Encouraging the use of recycled, reused, and bio-based materials through financial incentives and policy support.

Once this framework is in place, third-party verification of work site processes and emissions shall guarantee compliance with sustainability criteria during construction.

Furthermore, regulatory measures focused on the design and conception of infrastructure will help maintain long-term compliance throughout the infrastructure's lifespan.

By implementing these initiatives and criteria, we can create a robust foundation for resilient and sustainable infrastructure development.

2.1.1 Updating Sustainability Assessment

Every construction project involves several indicators that can be assessed from a sustainability perspective. These indicators traditionally play a crucial role in providing a general overview of the project's sustainability performance.

However, as public policies evolve to meet new resilience requirements, the relative weight of these indicators should also be reconsidered, as illustrated in Fig. 1 [1].

Standardizing requirements from a policy perspective is essential. To effectively evaluate a project's sustainability performance, it is thus imperative to define adequate criteria based on quantifiable factors, qualitatively ascertainable criteria, regulatory frameworks, and resilience requirements for infrastructure included in tender specifications.

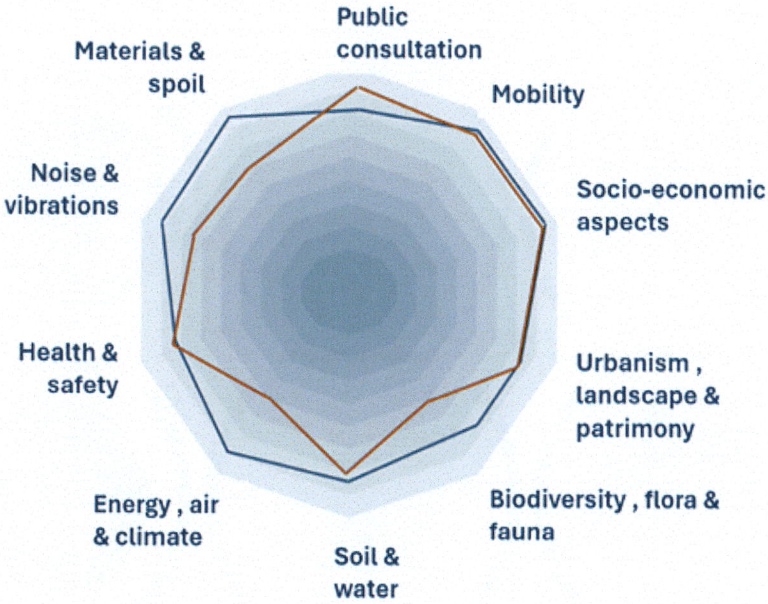

Fig. 1 Sustainability indicators relative weight drift between standard (orange) and resilient (blue) metro projects decision-making [2]

2.1.2 Behavioural and Policy Change

To fully implement these initiatives, a change in mindset is required to encourage a policy transition. Currently, decision-makers in the construction industry are typically manufacturers, owners, and developers.

However, we are advocating for a shift in mentality so that governments take the lead in promoting resilience and become key decision-makers when it comes to construction projects. This can be achieved by regulating the construction frameworks. Once this shift is accomplished, designers can better guide policymakers in implementing a new, more resilient approach to construction (see Fig. 2).

In this redefined scenario, designers will need to assume new responsibilities, including:

- Design for longer life spans;
- Increasing research and education in decarbonization strategies;
- Digitalizing design processes to ensure optimization from the early stages of the project (for example, Building Information Modelling or BIM);
- Designing based on the reuse of local materials, such as treating muck for concrete production;
- Collaborating closely with manufacturers to enhance the use of circular materials;
- Improving the management of on-site construction with circular design principles; and,

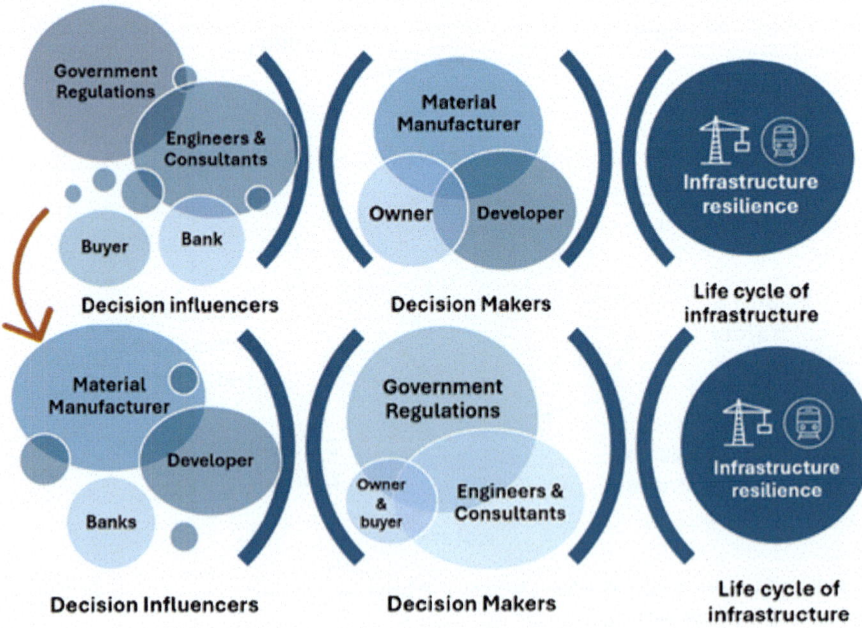

Fig. 2 Shift in decision-making intervenors on infrastructure life cycle [3, 4]

- Designing for disassembly and reuse when possible, including the reuse of Tunnel Boring Machines (TBMs) and construction facilities.

3 Exploring the Underground Potential

3.1 Geothermal Activation of Metro

Geothermal activation focuses on three key processes: extracting heat from the ground to heat buildings, monitoring and managing the changes in ground temperature, and injecting heat into the ground to cool buildings.

Geothermal activation of underground infrastructure is envisioned as a shortcut to ensure the self-sustainability of new metro networks. The feasibility studies for geothermal activation of subway projects typically involve the following steps:

- Identification of the horizontal and vertical alignment of the layout;
- Reviewing and defining existing geothermal solutions that can be applied to the project;
- Division of the line into—geological and/or hydrogeological—homogeneous sections;
- Application of geothermic abacuses to each identified homogeneous section to evaluate the winter and summer geothermal potential and the total thermal power that can be exchanged with the ground;
- Identification of potential receptors for geothermal energy;
- Evaluating the economic viability, regulatory requirements, and technical feasibility of the project.
- Outlining the necessary studies to be conducted in the subsequent phases of the project and planning accordingly.

Amberg Engineering has explored this methodology in collaboration with GEOEG—a leading service provider for the design, implementation, and optimization of energy geostructures—for Brussels Metro and also with construction engineering company NGE for L15 line of the Grand Paris project [5, 6].

3.1.1 Grand Paris Ligne 15

In the context of the opportunity study, an assessment of the geothermal potential of retaining walls and tunnels on Line 15 Ouest (L15O) has been conducted. The results indicate that geothermal activation of the line could mobilize significant thermal power, contributing to the sustainability and energy efficiency of the metro network.

By leveraging geothermal activation, underground infrastructure projects can achieve enhanced self-sustainability and energy efficiency, paving the way for more resilient and sustainable urban environments (see Table 1).

Table 1 Table of exchangeable power results for L15O [6]

Season	Tunnel (MWh)	Stations (MWh)	Total (MWh)
Summer	2,57	0,99	3,56
Winter	5,43	0,82	6,25

Table 2 Table of exchangeable power requirements per station for L15O [6]

In winter	In summer	All the year
162 kWh/station	376 kWh/station	376 kWh/technical rooms
In winter	In summer	All the year

A preliminary analysis of potential energy needs per station was also conducted. It is important to note that in the anticipated scenario, the energy demand for cooling is greater than the demand for heating. This increased cooling demand is primarily due to the continuous need for cooling the technical rooms.

The requirements per station would therefore be as follows (see Table 2).

The conclusion drawn from the preliminary analysis is that activating the retaining walls alone is sufficient to meet the year-round heating and cooling needs of the stations. This includes maintaining a continuous cooling system for the technical rooms.

However, there is a risk associated with using the system continuously in a single mode (cooling). This could lead to an annual thermal drift, resulting in a decrease in thermal performance over time due to the heat saturation of the ground. Therefore, further study on this specific issue is necessary to ensure long-term efficiency and sustainability.

3.1.2 Metro Nord Brussels

A similar study was conducted for the Brussels Metro, which included geothermal activation of the station slabs. The results indicated that the power generated from this method would be sufficient to provide energy not only for the stations themselves but also for energy-efficient housing within a 100-m radius of the metro stations (see Fig. 3).

3.2 Decarbonizing Construction Materials: Challenges and Opportunities

Current construction materials predominantly rely on energy-intensive, mineral-based extractive processes that cause significant environmental impacts throughout their lifecycle. Concrete and cement alone contribute approximately 1.5% of global CO_2 emissions [3]. Moreover, the prevailing 'take-make-waste' model in material

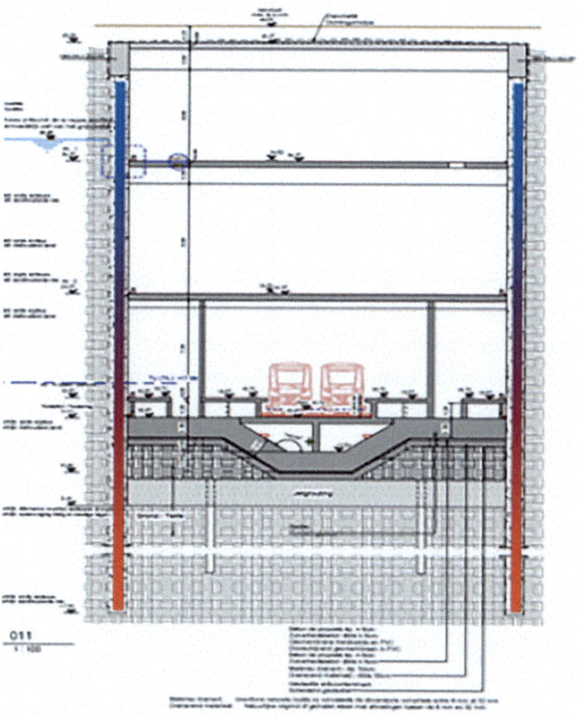

Fig. 3 Geothermic activation of Metro Nord Bruxelles station [2]

production means that most construction materials become waste at the end of their lifespan.

To decarbonize materials on a global scale within the construction sector, a fundamental shift in mindset is required to overcome resistance. Policymakers must engage stakeholders across the entire value chain to achieve this goal.

Investing in further research on materials should be a top priority for those within the material production chain. Recent studies have demonstrated that incorporating limestone fines into a CEM I (High-strength pure Portland cement for concretes requiring the highest strengths) and GGBS (Ground Granulated Blast-furnace Slag, cementitious material used as a concrete additive) ternary system improves material efficiency. Limestone fines can replace portions of both CEM I and GGBS, resulting in concrete with lower embodied carbon and reduced GGBS usage, while maintaining equivalent performance [7].

Policymakers must take decisive action to implement regulatory frameworks that impose new carbon footprint limitations on construction materials. Such restrictive policy changes would drive increased investment in alternative solutions [8]. The UK's update to the standard BS 8500:2023 serves as a notable example, incorporating the use of multi-component cement [9, 10]. This update reduces the proportion of Portland cement required for a given volume, thereby lowering the embodied carbon in the concrete produced.

By adopting these strategies, the construction sector can significantly reduce its environmental impact, contributing to a more sustainable future.

3.3 Harnessing Artificial Intelligence for Underground Space Engineering

The pursuit of innovative solutions to reduce carbon footprint has prompted a shift towards out-of-the-box thinking in recent times. Conventional approaches may no longer suffice in our quest for resilience, necessitating the exploration of creative solutions and a rethinking of our development approach. Artificial Intelligence (AI) emerges as a powerful tool with the potential to revolutionize geotechnical engineering, offering numerous possibilities for achieving this objective:

- Increased Efficiency: AI enables faster data analysis, enhancing efficiency in geotechnical engineering processes.
- Reduced Material Consumption: By ensuring more accurate designs, AI helps minimize material consumption, contributing to sustainability efforts.
- Enhanced Safety: AI facilitates a more controlled construction environment, thereby improving safety measures.
- Digitalization of Project Lifecycle: The integration of AI digitizes the entire project lifecycle, enabling data-driven decision-making processes that are both quicker and more efficient.

Furthermore, AI, as a data-driven tool, allows for the improvement of geotechnical models by using historical data combined with real-time analysis. This methodology enables rapid comparison of monitoring data with geotechnical information from surveys, resulting in a more accurate representation of soil behaviour.

The integration of machine learning algorithms into geotechnical engineering poses significant challenges but also offers immense potential for efficiency and safety improvements during construction. Real-time analysis of excavation data optimizes tunnelling strategies and extends the operational lifespan of machinery, ensuring greater efficiency and sustainability in construction projects.

4 Reimagining Underground Spaces: Madrid's Urban Transformation

Madrid's M30 road, a 32.5 km urban motorway, is undergoing a transformative reconditioning project. A recent urban proposal aims to transition a stretch of the M30 from an Open Cut to a Cut & Cover solution, creating a green recreational area above the current traffic lanes. This initiative not only beautifies the landscape but

Fig. 4 Current situation at Las Ventas bridge, Madrid [2]

Fig. 5 Foreseen urban action Las Ventas bridge, Madrid [11]

also reduces sound disturbances from heavy traffic, fostering better communication between neighbouring districts and revitalizing urban fabric perspectives.

While underground works initiatives are often hindered by high-cost estimations, mixed solutions like Cut & Covers present affordable opportunities to relieve urban pressure. Creative technical teams supporting policy decision-making can explore alternative ways to reinvent saturated metropolitan areas, fostering sustainable urban development within realistic budgets (see Figs. 4 and 5).

5 Conclusion: Towards a Sustainable Future

Systematic analysis of contextual constraints, coupled with well-planned implementation of innovative strategies like AI, are crucial steps towards a more efficient and sustainable future in construction. Underground infrastructure innovation opens up possibilities for exploring urban fabric in unprecedented ways, challenging traditional concepts of geotechnical and tunnelling engineering.

As we strive to achieve a better balance between infrastructure development and resilience, it becomes our generation's responsibility to take up this challenge. Engaging as engineers in this urgent request implies committing to the responsibility of changing direction and making sustainable urban development a reality. The future of our cities rests in our hands, and it is up to us to take action and bring about meaningful change.

References

1. Workshop on sustainability for Brussels Metro BMN—Amberg Engineering
2. Digital Madrid's Journal Gacetín Madrid https://gacetinmadrid.com/
3. United Nations Environment Programme (2023) Building materials and the climate: constructing a new future. Nairobi
4. United Nations Environment Programme (2022) Global status report for buildings and construction: towards a zero-emission, efficient and resilient buildings and construction sector. Nairobi
5. GEOEG (2020) Bruxelles Métro Nord: activation thermique des tunnels Analyse d'opportunité. Brussels Métro Nord—GEOEG for Amberg Engineering
6. GEOEG (2023) Ligne 15 Ouest du Grand Paris Express Analyse d'opportunité. Grand Paris Express, Ligne 15 Ouest—GEOEG for Amberg Engineering
7. German Tunnelling Committee (ITA-AITES) (2023) Sustainability in underground construction basics, boundary conditions and objectives DAUB-working group, Deutscher Ausschuss für unterirdisches Bauen e. V. (DAUB)
8. European Union (2023) The green deal industrial plan, Speeding up the contribution of Europe's innovative clean tech industries to net-zero. Luxembourg
9. MPA The Concrete Center (2023) BS 8500-1:2023 concrete—complementary British standard to BS EN 206—Part 1: method of specifying and guidance for the specifier, from Concrete Futures, Spring 2023
10. MPA The concrete center (2023) BS 8500-2:2023 concrete—complementary British standard to BS EN 206—Part 2: specification for constituent materials and concrete, from concrete futures, Spring 2023
11. El Economista (2023) Underground Works on M-30 at Ventas beginning by the end of the year https://www.eleconomista.es

Underground Metro for Modern and Medium-Sized City—A Case of Copenhagen

Tommy Olsen

Abstract In 1992, the Danish parliament decided to construct a new light driverless metro in Copenhagen, with two lines stretching 20 kms across twenty-two stations. Funding for the new metro system was obtained through the sale of municipally owned land. An underground metro was selected over a surface-level tram network because it offered higher speeds, reduced visual and noise pollution and lower incidence of accidents. Copenhagen has buildings from medieval times that are sensitive and of significant heritage value. An underground metro helps maintain the look and identity of the city. This chapter describes how the layout of the stations considers climate change mitigations, such as flood risk from sea level rise and increased rainfall, and highlights innovative ideas considered to reduce the construction's carbon footprint and improve biodiversity in the districts influenced by the new metro line. This chapter is to a large degree based on articles from the book "Copenhagen Metro—Copenhagen Metro Inauguration Seminar", 21–22 November 2002 in Copenhagen, published by COWI A/S/ [1]. It shows how a vision was developed for the first line of the Copenhagen metro and how this vision has been amended as the demand for sustainable development increased over the years.

1 Historic Background

The Danish capital of Copenhagen is located on the western coast of the Øresund Strait, separating the Nordic countries of Denmark and Sweden at the entrance to the Baltic Sea. Also known as "merchant port", the city was founded in 1160, and its strategic location made the capital city an ideal place for merchants to meet and trade. Understanding the local context and the past is fundamental when developing change for the future. In the late 1800s, due to the industrialization of society, many people in Denmark moved from rural to urban areas. Like many cities in Europe, Copenhagen experienced a significant increase in the number of inhabitants, and

T. Olsen (✉)
COWI A/S, Parallelvej 2, 2800 Kongens Lyngby, Denmark
e-mail: tool@cowi.com

© The Author(s), under exclusive license to Springer Nature Singapore Pte Ltd. 2025

M. Agrawal (ed.), *Underground Spaces for Climate Resilience and Sustainability*,
Advances in 21st Century Human Settlements,
https://doi.org/10.1007/978-981-96-0547-7_11

new districts were built around the medieval and historic centre. A new harbour area, Nordre Frihavn or "Northern Free Harbour", was constructed to accommodate larger ships and compete with the nearby Hamburg Harbour after the completion of the Kieler canal, which eased the passage of merchant ships between the English Channel and the Baltic Sea. A new shipyard, for the company Burmeister and Wain, in the area of Refshaleøen, was modelled after the shipyard in Liverpool and soon became the largest industrial working place in Copenhagen.

The need for transporting people around the city increased and the nature of public transportation in the city evolved from horse-driven carriages to buses and later trams. A central stretch of rail line was placed underground between Vesterport and Østerport stations, in the Boulevardbanen open in 1917. The rail and underground rail line followed the former moat and rampart system that protected the inner city. The tunnel section of 1.5 km was the only underground transportation infrastructure until Copenhagen metro lines M1 and M2 started operating in 2002.

In the 1930s, the electric S-train system began operating across a 12-km stretch from Hellerup to Valby, using the tunnel from Boulevardbanen. The S-train is a public transportation constructed at grade to connect the outer suburbs with the city centre; the system was expanded between 1940 and 1980 when most of the 170-km length of S-train lines were constructed. During the 1950s and 1960s, new towns grew with S-train and motorways connecting different parts of Copenhagen. The city centre became congested with cars taking up significant space. City planners identified the problem and politicians did something about it, as cycle paths along the major roads to and from the city were established as alternative modes of transportation to buses and cars. In the city centre, cars were removed from Strøget, a kilometre long street which was the biggest shopping street for pedestrians in the world when it opened.

During the 1970s and 1980s, unemployment in Copenhagen grew and housing availability in the city struggled to keep up with the growing demand. People moved from Copenhagen to city suburbs, where better and more affordable housing options were available. The situation in Copenhagen was not sustainable, and something had to be done. The solution found was to complete the motorway network in Denmark, connecting all the major cities from east to west and from north to south. Additionally, a major fixed link across the Great Belt was planned to connect the island of Zealand and Copenhagen with the island of Funen, extending to mainland Jutland and thus physically connecting Zealand to Germany. This fixed link was to include both road and rail connections, despite the business case for the rail component being less robust than that for the road.

Shortly thereafter, the decision was made to construct the Øresund link, connecting Malmö in Sweden with Copenhagen. During the 1990s, shipyards in Malmö (Kockums) and Copenhagen (Burmeister and Wain) faced severe global competition, which ultimately led to their closure after many years as major employers in the region.

Close to the Øresund Bridge and Copenhagen Airport, on the island of Amager, south of Copenhagen, lay a large abandoned military area called Amager Fælled. City planners and politicians envisioned there a new town with housing and workplaces, naming it Ørestad. To enhance its appeal, Copenhagen's first metro line

Fig. 1 Construction of the Copenhagen Underground railway section Boulevardbanen. Photograph from 1914. Public domain [2].

would connect Ørestad to the inner city. Positioned centrally in the Øresund region with excellent transport links, Ørestad was an ideal location for development. The financing model for the metro involved selling the unused military land owned by the state, and the anticipated increase in land value driven by the new metro line would fund the metro's construction.

Effective transportation network adds value to land and to society. More underground metro networks are being introduced or expanded in cities around the world to respond to the need for more sustainable forms of transportation. To develop plans and designs for growing sustainable cities that provide long-term benefits for future generations, one needs to know about the people, their history and the identity of the place (see Fig. 1).

2 Project Vision, Objective, and Change

From the very beginning, the project had a strong and clear vision. Although the concept of "sustainability" was not explicitly used as a guiding principle in the development of Copenhagen's first metro line, many elements that we now consider more sustainable were incorporated into the project vision and the vision later used as basis for the design requirements. The organization established to develop the new town in Ørestad and the new metro line, called Ørestad Development Corporation, /1/, page 15, articulated the following vision for efficient and easy travel of users:

- Travel time is short
- Reliability is high
- Information is fast, relevant, and reliable
- The customer receives a friendly and effective service
- The customer feels safe
- The customer experiences the journey as clean and comfortable
- The customer experiences that the Metro "listens to them"

A design guide to support the development was made by the architects and engineers to reach a result that would deem important the aspects defined by the project, organized in such a way that members from different fields of expertise could meet, discuss and agree on the direction for various general and specific topics. With clearly stated objectives, collaboration was made easier as the discussions would be focused on solutions rather than on the objectives. Experience from the first metro lines M1 and M2 shows that, with a known objective and clear targets, it is easier to reach a goal and a vision in a collaborative approach, as opposed to projects with a moving target.

2.1 Open Stations and Tunnel Alignment

Two key features of the Copenhagen metro are user-friendly and inviting metro stations. These features have been instrumental in defining that the city's underground station platforms should receive ample daylight and be easily accessible from ground level, and the locations of the metro stations were strategically selected to maximize passenger numbers. Since the first metro line aimed to attract new customers unfamiliar with underground urban travel, the objective was to make the metro as appealing as possible as an alternative to cars and buses. Consequently, metro stations were often situated close to, or integrated within, existing spaces such as squares or parks, providing easy access from multiple directions. The design development considered the entire journey from door to door and not just the experience at the stations and during transit. This holistic approach included ensuring safe and pleasant roads or pathways from the journey's start to its end, providing shopping opportunities along the way, and incorporating underground cycle parking connected to the stations. These features were carefully considered in the station design to enhance the overall travel experience.

The focus on construction costs and minimum impact on neighbours from construction activities meant that the stations were designed to be no larger than necessary. The footprint area of the island platforms and the rails on both sides defined the perimeter of the deep excavations. Space above the metro rail line was allocated for technical rooms, and access to the stations at the concourse level included provisions for cycle parking. Access to the platforms was ensured by escalators, and openings were incorporated for daylight to reach the platforms, creating a pleasant experience for users, and avoiding the feeling of being trapped underground (see

Fig. 2). The layout of the stations was standardized across the network, fostering a sense of familiarity for passengers. The original concept was developed after a study tour of the Paris Metro (see Fig. 3).

Fig. 2 Metro station located at an existing square—here Kongens Nytorv. *Photo* Tommy Olsen

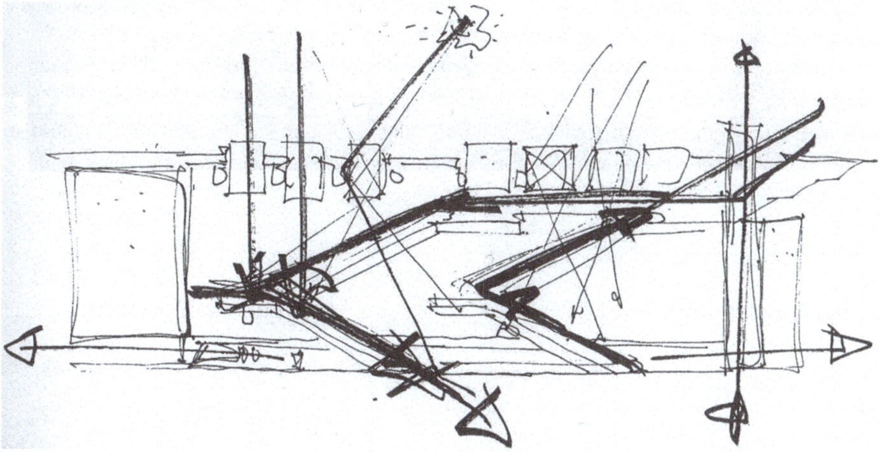

Fig. 3 Original Metro station concept developed on a napkin in a restaurant in Paris. *Photo* COWI A/S

For the later metro lines, the colours and materials used in the stations reflect the characteristics of the surrounding area. For example, red tiles might be used to mirror nearby buildings, or green walls to allude to the nearby avenue of trees adjacent to the station. This approach helps integrate the stations into their local environments while maintaining a cohesive design language. Most of the underground stations were constructed using deep and open excavation technologies. Given the region's unfamiliarity with deep excavation techniques, foreign contractors were invited to bring in technology not commonly used in Denmark.

Deep excavations were carried out using reinforced secant-pile walls. The piles were installed using boring rigs, with the pile-toe placed some metres into the hard limestone. Due to the sensitivity of the centuries-old buildings in the centre of Copenhagen, many of which are over 200 years old, lowering the groundwater table would be unacceptable. Many of the old buildings in Copenhagen have foundations made of wooden piles which are preserved by being fully submerged in water and protected from deterioration due to the lack of oxygen. Lowering the groundwater table around these wooden piles would risk exposing them to air, leading to potential damage. Therefore, many of the deep excavations were equipped with re-infiltration wells to ensure a stable groundwater table, despite the deep excavation inside the retaining wall.

The retaining walls were temporarily supported by steel struts. In the permanent structure, the secant piles were supported by in-situ concrete slabs and struts. The secant piles are designed to support earth pressure in the permanent case, whereas an inner lining is designed and constructed to withstand water pressure on the station. To mitigate the risk associated with tunnel construction, the alignment of the metro between the stations was chosen to follow the limestone layer. Bored tunnel construction technologies were new to Denmark in 1990, so a solution with a reasonable level of risk was selected. Construction of more recent metro lines with new technology and methodologies has demonstrated that bored tunnels can be constructed in shallower locations within sand and clay formations above the limestone, reducing the need for limited resources, including construction materials.

It is clear from the design of the stations and tunnel alignment today that the vision of the first metro lines helped in design decisions, both in the development of the design concept—for example, with skylights reaching the platforms—and in the selection of the finishing work, for example, the architectural cladding for the stations, alluding to the location of each station (see Fig. 4).

3 Metro Design for Climate Resilience and Flood Response

As mentioned earlier in this chapter, Copenhagen is a city situated along the coastline, and its growth has often involved reclaiming land in the water. The city's economy has historically thrived on trade, frequently conducted via sea transport. Over the

Fig. 4 Metro station in Frederiksberg Alle, famous for its large (green) Linden trees. *Photo* COWI

years, Copenhagen has expanded and evolved within the constraints and opportunities presented by its coastal location, shaping its development as a vibrant coastal city.

Climate resilience and flood response were incorporated in the design of the first metro lines in the 1990s, with protection against heavy rainfall. Unusual to local engineers at the time, it followed the same risk approach used to design against extreme water levels, something local engineers where more familiar with, since Denmark is a flat and coastal country.

Data box: Resilience as defined by the United Nations: "The ability of a system, community or society exposed to hazards to resist, absorb, accommodate, adapt to, transform and recover from the effects of a hazard in a timely and efficient manner, including through the preservation and restoration of its essential basic structures and functions through risk management."

Denmark is situated far from regions with significant seismic activity such as earthquakes, tsunamis and volcanoes; the nearest such areas are Iceland, Italy and Greece. Earthquakes above a magnitude of 4 are rarely recorded in Denmark, so seismic events are not typically viewed as a significant risk. Although southern

Scandinavia does not experience major typhoons, in the last 10–15 years there has been an increase in both total monthly and yearly rainfall, as well as in rainfall intensity. Historically, Denmark has not been prone to natural disasters, so the design of large infrastructure projects has not traditionally taken major disastrous events into consideration. However, with the onset of climate change due to global warming, there has been a shift in focus towards extreme weather events. Weather conditions were not given the same level of consideration in the 1990s as today, reflecting the evolving understanding of climate change and its implications.

After a major stormwater event in Copenhagen in the summer of 2011, where the old and combined rain and sewage system was overloaded in some areas of the city, a stormwater strategy for Copenhagen was developed, including plans for the construction of several stormwater tunnels. Fortunately, during the 2011 event, the flooding of low-lying areas in Copenhagen did not affect the metro. The increased focus on heavy rainfall in the Copenhagen area has led to a heightened awareness of the risk of flooding. Areas vulnerable to flooding have been identified, and measures have been taken to mitigate the risk around entry points to the metro, such as elevating the entry points or installing barriers to divert water away from the metro entrances.

Climate adaptation and the potential impact of global sea level rise have been taken into consideration since the planning of the first metro line in the 1990s. Design levels for both extreme seawater, groundwater and severe rainwater events in the future were established as part of the design basis. As a starting point, the metro was designed against a rainfall event with a 2000-year return period and an extreme water level event with a 10,000-year return period. Water levels were selected in a manner that ensured the structural integrity of the metro system was not entirely reliant on the accuracy of actual water levels. Watertight doors were used for the mechanical and electrical installation rooms to mitigate the risk. Some tolerance for water ingress during very rare events was accepted. Moreover, the ground level at all entry points to the metro was positioned above the worst-case scenario of raised groundwater levels, further minimizing the risk of water infiltration into the metro system.

The performance of the metro lines during the last 25 years shows that climate resilience and flood response can be incorporated into the design criteria within construction cost budgets and with an acceptable outcome (see Fig. 5).

3.1 Sustainability as Design Criteria

The tunnelling industry sees the subject of sustainability becoming increasingly more central in the development of major projects, in the way visions are being developed and—consequently—in how project requirements are being defined. A possible new line for Copenhagen, M5 is being developed, aiming to move the project towards "more sustainability" compared to previous lines. Initially, the United Nations Sustainability Development Goals (UN SDG) were used as a source of inspiration. However, once the potential impacts and benefits were better understood, Metroselskabets defined project-specific sustainability criteria. More specifically,

Fig. 5 Construction of metro station in deep excavations. *Photo* COWI

Metroselskabet has defined a "Value Key" with four key areas to guide the project, as described further in the text.

Until around 2020, sustainability was not widely integrated into underground infrastructure projects in Denmark. However, in the building industry for housing and offices in Copenhagen, owners, planners, designers and contractors have been implementing sustainability practices for some years. Large under- and aboveground infrastructure projects typically have a larger climate footprint as individual projects than single-building projects for housing and offices. Since the energy crisis in the early 1970s, when the country was dependent on the import of fossil-fuelled energy, the regulations for getting building permissions for housing and industry have focused on reducing energy consumption. In addition, the focus has been broadened to embodied carbon from building materials. As a consequence, housing has a framework of standards, norms and legislation where sustainability requirements can be incorporated, and certifications are common—and perceived as good investments—in building projects.

Major infrastructure projects, like underground metros, normally have individual bill (construction act), and there is no shared legal framework within the infrastructure industry to anchor sustainability requirements. Based on objectives defined in the early stages of an infrastructure project (like tunnels), impacts from sustainability requirements should be investigated in the concepts and preliminary design phases to better understand the detailed implications (for example cost, time, risk, social and environmental impact) in case these requirements are to be fulfilled by the contractors if implemented in the procurement specifications.

Awareness of sustainability issues has been growing in the country, and politicians and project owners increasingly recognise their role, responsibility and impact on decisions related to sustainable development. This growing awareness is reflected in a broader recognition of the importance of environmental responsibility in shaping the future of urban development. For the past years, the UN Sustainable Development Goals (SDGs) have been used in Denmark to help identify and communicate impact, risk, and mitigations at a strategic level. For example, UN SDG no. 13—"Climate action" as a strategic goal—is often used on tunnel projects where the focus is on sustainability. This goal can be materialised (1) as a CO_2-e target, in terms of actual tonnes of reduced CO_2-e, (2) as a percentage of reduction compared to a baseline or (3) more specifically, assigning low-carbon material and construction equipment powered by renewable energy. The introduction of the SDG's has had a positive effect on integrating sustainability in e.g. construction projects. However, there also seems to be a move from a stringent use of the goals to a more selective approach once companies have established their main positive and negative impacts.

Legislation derived nationally or from the EU, for example on protecting nature and biodiversity establishes several guidelines and rules for sustainable development, design, and operation of infrastructure projects. Innovation is needed to "move the needle", where the cost of construction is kept at an acceptable level in combination with minimizing the environmental impact (see Fig. 6). Without innovation, and with rigid rules and regulations based on past technology, it is more difficult to make significant changes. In the planning and design phase of M5, Metroselskabet established an Innovation Programme, with the purpose of exploring how technical innovation can support achieving the project sustainability criteria. To date, numerous projects have been funded within this context, and the programme is now a permanent team within the organization.

Sustainable development measures are often perceived as an additional cost to a project, and it is crucial to distinguish between what is truly necessary and what is optional. Sustainable benefits can influence the business case of a project and must be clearly communicated, as it is not feasible to implement every sustainable measure due to budget constraints. Aiming at reducing material quantities and climate footprint by optimizing the design could enable a reduction in the overall cost, but this also depends on factors other than quantities and used materials, such as construction time, complexity of construction, availability of equipment, labour, and risks. In any case, the impact of design and construction needs to be monitored continually to support good decisions.

In Denmark, the use of a framework for sustainability certification of infrastructure projects is not common. Yet, a structured and systematic approach when working with sustainability benefits is of great value to any infrastructure project.

Focus on future-proofing and design flexibility can also drive greater attention to sustainability. However, it is challenging to design against unforeseen circumstances, especially in projects with tight budgets. Experience from the performance of past projects should be taken into account when planning for the future, as they host many lessons to be learned.

Fig. 6 Entrance to metro station Havneholmen, line M4, open June 2024. *Photo* Tommy Olsen

When it comes to whole-life cost and asset performance, there is a strong emphasis on flexibility. Metro infrastructure typically has a long service life, but mechanical and electrical installations may need to be replaced multiple times during this period. Therefore, ensuring flexibility in design is critical to accommodate future upgrades and changes while optimizing the performance and longevity of the infrastructure.

Like climate impact, the impact on biodiversity is an emerging focus area for construction. Consideration should be given to preserving and enhancing biodiversity in areas affected by construction and operation. Methods to improve biodiversity for a metro project can be to choose the right plants and trees suitable for the location and ensuring living spaces for insects. Social aspects, such as accessibility and equity, are always taken into account in metro planning. A well-functioning easily accessible, public transport system is considered essential in a welfare society like Denmark. This was reflected in the customer vision developed for Ørestaden in the 1990s, and in the metro lines M1/M2.

4 New Metro Lines

The first metro lines in Copenhagen—M1 and M2—opened in 2002, with M1 connecting the new development area of Ørestaden with the inner city, and M2 extending to the island of Amager, as far as Lergravsparken. Shortly after, M2 was extended to Copenhagen Airport and M1 and M2 were extended west to Vanlose. Both lines featured underground, ground-level, and elevated sections (see Fig. 7).

In 2019, a significant extension was completed with the opening of the M3 Cityringen, or Circle Line, which connected all districts around the inner city to each other and to the existing metro lines and regional S-train system.

In 2020 the M4 line was opened, partly sharing tracks with the M3 Cityringen, to provide connectivity to the new urban development of Nordhavn. With Ørestad fully established, Copenhagen looked for new areas for development, focusing on housing and business opportunities. The financial model used for Ørestad is still being applied to new developments: land owned by the state and municipality is sold to developers for housing and office projects, with the guarantee that the metro will be connected to the new area.

The people and politicians of Copenhagen are highly satisfied with the metro service and are eager to expand the network to make it even more attractive for residents and visitors. Like many other major cities worldwide, Copenhagen grapples with significant road congestion, noise, and pollution issues. A robust public transport

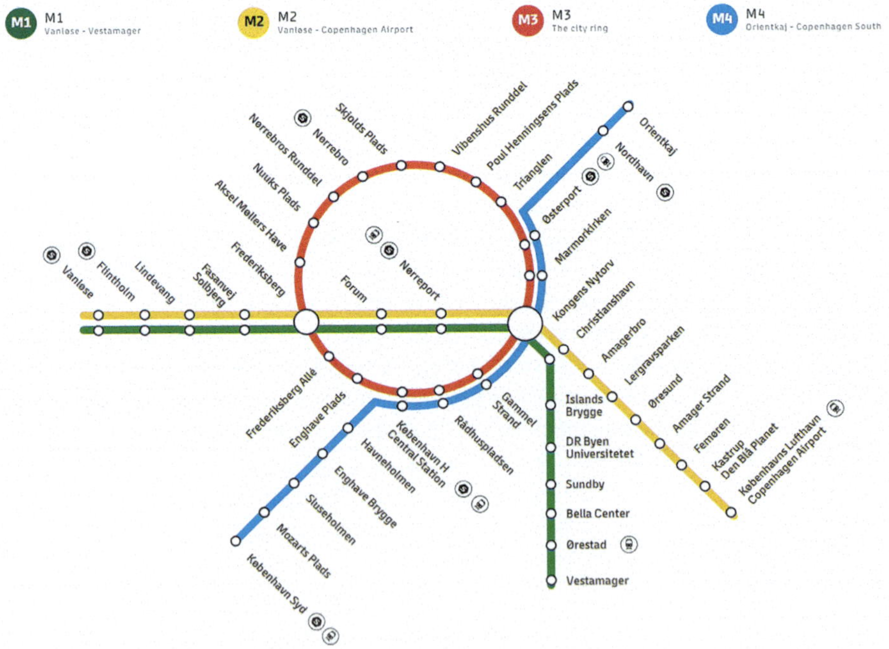

Fig. 7 Metro lines, Summer 2024. *Image source* Metroselskabet

system and an extensive network of cycle paths throughout the city play crucial roles in alleviating these negative aspects of life in the city. The south branch of M4, which began operating in the summer of 2024, provides five new underground metro stations connecting the newly developed urban district of Kalvebod Brygge and the new and large transportation hub "Copenhagen South" to M3 Cityringen and inner city via the old quarters of Sydhavnen district.

Plans are underway for a possible new metro line, M5, which should provide a metro for the future island of Lynetteholmen and with the possibility of making a better connection between the southern part of Copenhagen on the island of Amager and the rest of the city (see Fig. 8). Given the projected growth in housing developments, traffic studies suggest that the capacity of metro line M1 will be reached by the 2030s, further necessitating the planning and construction of M5, including its new additional harbour crossing to meet future demands. To ensure the UN SDG is translated into decision support, a so-called value-key was developed and clarifies the objectives of the project in addition to meeting the legal requirements and design codes. It defines the targets more specifically on how to measure and document the improved performance of the solutions. A performance baseline of the selected parameters was established based on solutions from the previous metro lines and is used to demonstrate improvements and the value of innovative alternatives.

The value key assesses the benefit of various solutions and alternatives. It also helps in the assessment of new technologies which may be beneficial to such projects. If the new solution does not provide additional benefits, namely more value compared to a conventional method, then it is not investigated further.

The main focus areas on sustainability for the construction of new metro lines in Copenhagen are listed below. In brackets are examples of how to meet the specific objective shown in the value key.

(a) Resource and climate impact (minimize use of energy, use less reinforced concrete and steel, use low carbon materials, use equipment powered by renewable energy, recycle more, excavate and transport less).
(b) Passenger experience and operational robustness (improve passenger experience both in locations and layout of metro stations, integrate and connect systems, minimize sensitivity from rare unfortunate events)
(c) Safety and Working environment for workers and users (develop safety culture further, work more standardized with safety, use more prefabrication and improve quality)
(d) Contributing to the City (better connections also with other types of sustainable transport systems, more biodiversity, more climate adaptation and resilience, contribute during construction).

The process serves as a valuable guide for owners, planners and designers, facilitating the achievement of common goals and improved performance. This method bears similarities to the development of the first metro lines in Copenhagen.

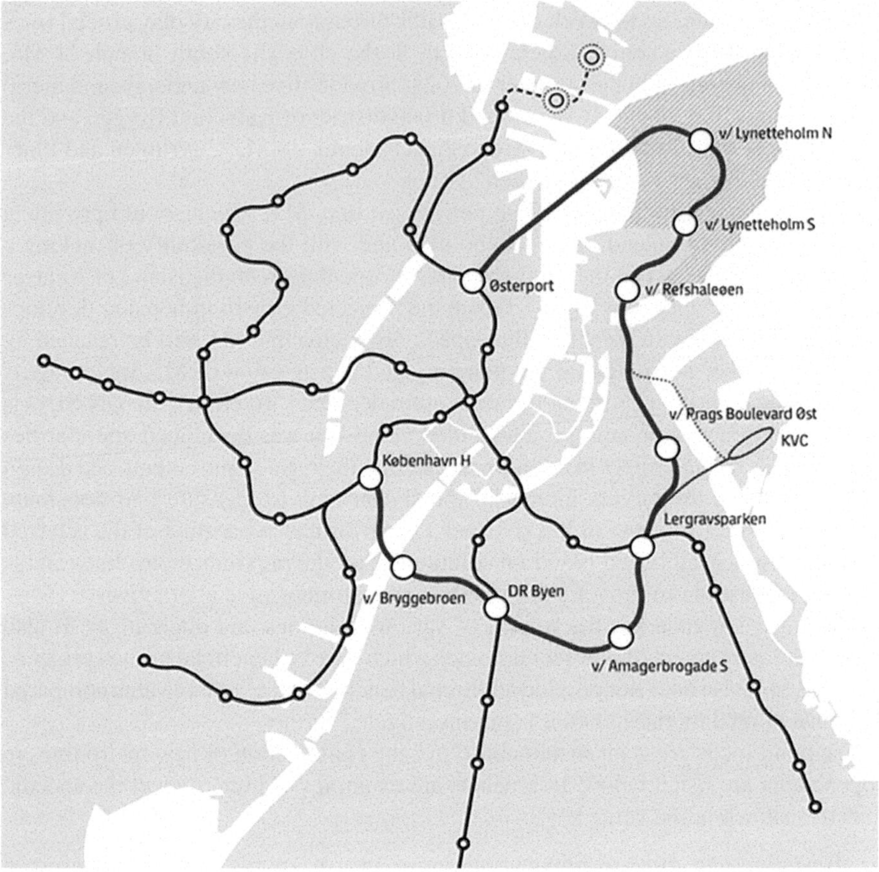

Fig. 8 New proposed metro lines. The EIA for M5 goes into public hearing in August–September 2024. *Image source* Metroselskabet

References

1. COWI A/S (2002) Copenhagen metro inauguration seminar 21–22 November 2002. Copenhagen, Denmark. Available via COWI A/S
2. Wikimedia Commons (1916) File: Boulevardbanen-Aug-2014, Source: The magazine "Før og Nu". Public domain

Enhancing Urban Resilience: Lake Area Metro Transforming Connectivity and Quality of Life

Hannes Strebel and Antonia Cornaro

Abstract The Lake Area Metro is a pioneering solution connecting the Lake Zurich region with the bustling city of Zurich, enhancing lakeside communities' quality of life. Historically, railway networks along the lake's shores fostered connectivity between villages without disrupting the landscape. However, the unregulated urbanisation of the twentieth century led to congestion and pollution. In response, "PLAN HECHT" was conceived, featuring a subterranean subway ring around the lower lake basin. This modern transportation infrastructure operates efficiently, embracing climate-resilient design principles and repurposing former S-Bahn routes into bike trails. Subway stations serve as vibrant hubs, enriching social interaction and commerce while promoting ecological restoration. The Lake Area Metro exemplifies holistic urban resilience, addressing transportation, environmental, and socio-economic dimensions, creating vibrant, liveable communities. The chapter will explore the project's rationale and ground-breaking concepts developed by students from ETH Zurich.

Keywords Landscape · Mobility · Vision · Metro · Underground · Connection · Community

1 Introduction

PLAN HECHT, a visionary traffic concept, is transforming mobility and enhancing the quality of life in the Lake Zurich region. A road tunnel beneath the lake alleviates traffic congestion, while the LAKE AREA METRO, with trains gliding underground,

---

H. Strebel
Dipl.-Ing. Architect HDK SIA, Wolfbachstrasse 15, 8032 Zurich, Switzerland
e-mail: hannes.strebel@mac.com

A. Cornaro (✉)
Urban Planner, Expert and Lecturer Underground Space, Amberg Engineering AG, Trockenloostrasse 15, 8105 Regensdorf-Watt, Switzerland
e-mail: acornaro@amberg.ch

© The Author(s), under exclusive license to Springer Nature Singapore Pte Ltd. 2025

M. Agrawal (ed.), *Underground Spaces for Climate Resilience and Sustainability*, Advances in 21st Century Human Settlements,
https://doi.org/10.1007/978-981-96-0547-7_12

connects the region with high-speed rail. This initiative eliminates traffic lanes that have long separated communities from the lake, expanding access to the waterfront and creating a unique local recreation area.

1.1 History of the Region

Prior to the late eighteenth century, the Lake Zurich region thrived on agriculture, viticulture, fishing, and crafts. Villages surrounded by agricultural meadows, fields, and vineyards defined the landscape. Transportation between these villages and larger towns primarily relied on horse-drawn carts or ox carts traversing country roads, while goods were ferried across the lake via barges.

The onset of industrialisation in the nineteenth century marked the establishment of manufactories and later, textile, chemical, and machine factories within the villages. However, the burgeoning industrial sector underscored the inadequacy of water transportation for rapid industrial growth.

In 1875, the Schweizerische Nordostbahn Gesellschaft (NOB) inaugurated the left-bank railway line, connecting Zurich's main station to Näfels via Ziegelbrücke. This was swiftly followed in 1894 by the right-bank line, extending from Zurich's main station through Meilen to Rapperswil. Concurrently, construction of the lakeside road commenced on both banks, with land reclamation from the lake facilitating the process and circumventing lengthy expropriation procedures.

Despite robust rail and road links to Zurich, a notable gap persisted between communities on the left and right banks, posing a significant logistical challenge and impeding seamless connectivity (see Fig. 1).

1.2 Current State of Affairs

Urban sprawl in Switzerland shows no signs of abating and has now encroached upon the shores of Lake Zurich.

A uniform urban sprawl is increasingly spreading across both shores of the lake. Both banks of the lake are witnessing a uniform expansion of urban development, transforming once picturesque lakeside villages into a monotonous dormitory town housing nearly 300,000 residents. The quaint village centres, once bustling with local stores, bakeries, inns, and businesses, have now dwindled, with only a few large-scale distributors catering to the population's needs. The vibrant village life, characterised by numerous clubs and societies, is visibly fading away.

The lakeside road, already stretched to its limits, struggles to accommodate the burgeoning commuter traffic, leading to traffic chaos and environmental degradation. Despite frequent S-Bahn services, the rail network is nearing its capacity, exacerbating the congestion woes. Furthermore, the bustling rail and road corridors sever the residents from lake access, exacerbating their plight. Although the lakeshores are

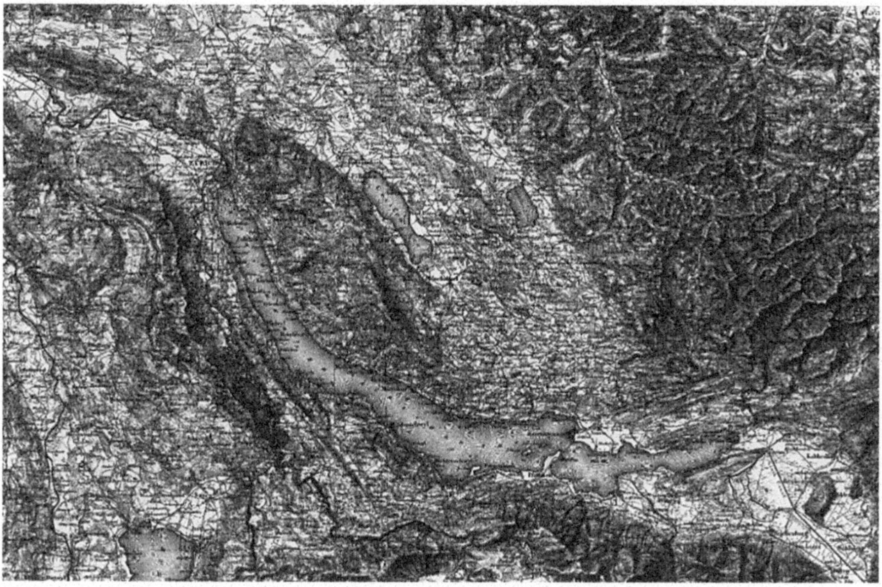

Fig. 1 The Dufour map of 1865 shows sparse settlement on both shores of Lake Zurich. The villages are only connected to the city of Zurich by a country road. Zurich had had a railroad station since 1847. Railroads ran from there to Baden (Spanisch Brötli Bahn). *Source* Swisstopo (Creative Commons. Link to source: https://www.swisstopo.admin.ch/en/dufour-map)

officially state-owned concession land, they are increasingly privatised, depriving the general public of recreational spaces.

The region grapples with a dearth of employment opportunities, and cultural, educational, and sports facilities. A significant portion of the population comprises retirees, while others commute to work in Zurich due to unaffordable housing. As night falls, commuters return to the dormitory-like agglomeration, detached from the vibrant city life.

Without immediate and drastic planning interventions, the region faces an imminent traffic collapse within the next two decades, exacerbating the existing challenges and compromising residents' quality of life.

1.3 Envisioning the Future

Given the looming traffic crisis, urgent and transformative measures are imperative for future planning:

- Eliminating the divide between the population and the lake by removing the railway and lakeside road barriers.
- Reclaiming and revitalising the lakeshore zone into a vibrant recreational area.

- Enhancing public and private transport connections between the lake communities and Zurich city centre while significantly reducing emissions from the transport infrastructure.

The Lake Zurich Master Plan 2050 (Plan Hecht), developed by architect and author Hannes Strebel, emerged as a visionary solution to address these challenges.

2 Plan Hecht

Introduced in 2014, the Plan Hecht garnered significant media attention and proposed the following key initiatives:

- Construction of an underground ring around Lake Zurich.
- Development of a motorway tunnel connecting Rüti ZH to Wollishofen.
- Establishment of a bike trail along the old railway line encircling Lake Zurich.
- Introduction of speedboat cross-connections between lakeside villages.
- Creation of a complete lakeside path encircling Lake Zurich.

By 2019, substantial progress had been made towards realising these goals, including the lowering of the railway line between Zurich Tiefenbrunnen and Thalwil and the establishment of public lakeside paths in Zurich, Küsnacht, and Wädenswil. These measures mark significant strides towards achieving a more connected, sustainable, and resilient future for the Lake Zurich region.

The Plan focuses on five measures to create a liveable and economically attractive future.

2.1 Measure 1: Reducing Surface Road Congestion via Pfannenstiel See Tunnel

This measure aims to alleviate congestion on Seestrasse and the A53 and A3 national highways near Rüti and Wollishofen by connecting them through the Pfannenstiel See Tunnel.

This tunnel will serve as a direct connection between these highways, diverting commuters and through traffic underground. As a result, the villages on the right bank of the lake will no longer be burdened by heavy traffic passing through their areas, leading to a significant reduction in congestion along the lakeside roads (see Figs. 2 and 3).

With the implementation of this tunnel, Seestrasse will be repurposed as a feeder road primarily for emergency and delivery services, further alleviating traffic pressure on the lakeside communities. Additionally, by relieving congestion and enhancing traffic flow, the tunnel project will pave the way for the ecological revitalisation of the

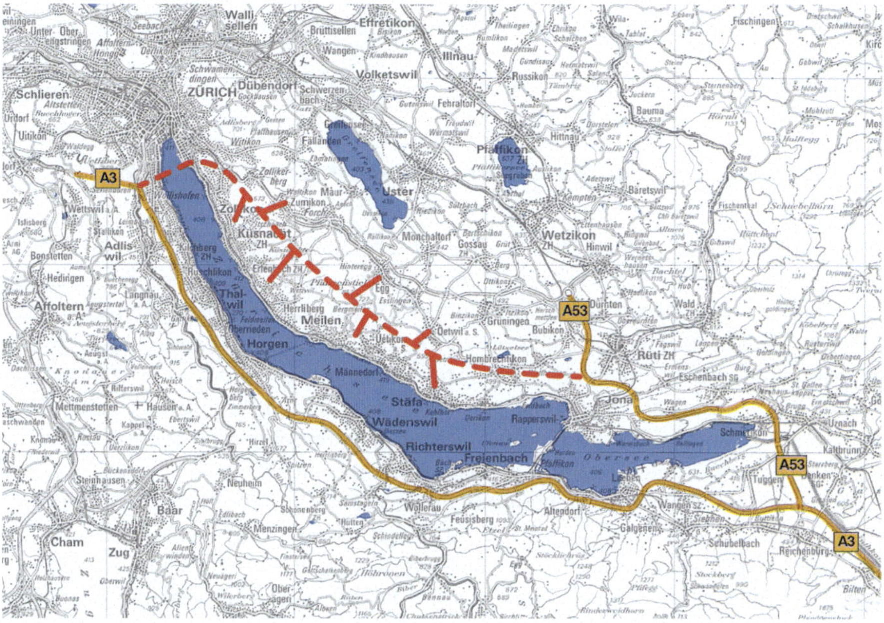

Fig. 2 A road tunnel between Rüti and Wollishofen relieves the Seestrasse and connects the villages on the right bank of the lake to the national road network. (Yellow—existing A3 highway or A53 national road; red—planned tunnel with feeder roads). *Image* Drawing Hannes Strebel

Fig. 3 By laying the road underground, the villages are freed from through traffic. The population will have long-awaited access to the lake. *Image* K. Jähne, Tunnel Leonberg Engelberg, creative commons

lakeshore, transforming it into a vibrant recreational area for residents and visitors alike.

2.2 Measure 2: Ecological Enhancement via Lakeside Path

Measure 2 seeks to address the ecological degradation of Lake Zurich and its surrounding areas by implementing an attractive lakeside path around Zurich Lake. Over the past two centuries, the ecosystem of Lake Zurich has suffered significant damage due to land reclamation and bank constructions, particularly during the nineteenth and twentieth centuries. These have led to the destruction of natural shallow water zones and spawning grounds, while the construction of bank walls has provided limited ecological value. Even today, ongoing construction activities continue to encroach upon natural habitats along the shoreline.

To counteract these effects and protect the local ecosystem, Measure 2 proposes the creation of reed beds along the lakeshore. These reed beds will serve as essential habitats for fish and amphibians, providing them with natural spawning grounds. Additionally, a lakeside path will offer the local communities a peaceful and uninterrupted walkway through a unique landscape.

Unlike traditional roads, this path will be free from interference by other road users, allowing visitors to enjoy the natural beauty of the lakeshore without obstacles or barriers. Overall, Measure 2 aims to enhance and protect the habitat for plants and animals while providing the community with a valuable recreational space accessible to all (see Fig. 4).

2.3 Measure 3: Metro Ring Around the Lake Area

Measure 3 introduces the Lake Area Metro, envisioned as a subway ring encircling the lake area, providing efficient and reliable transportation for residents and connecting them with key urban centres, educational institutions, and healthcare facilities.

By placing railroads underground in densely populated areas, this initiative aims to restore access to the lake for the local population, thereby revitalising its function as a recreational area and promoting community well-being (see Fig. 5).

The advantages of a subway compared to conventional rail connections are manifold:

- Efficiency: Underground trains are generally faster than conventional trains as they operate in dedicated tunnels, free from road traffic or other obstacles.
- Schedule reliability: Operating on exclusive tracks and in designated tunnels enhances reliability, with timetables less susceptible to disruptions caused by weather or traffic

Fig. 4 A promenade along the lake will significantly enhance the entire agglomeration. *Image* Hannes Strebel

- Higher capacity: Metros can run at higher frequencies than conventional trains, accommodating more passengers than conventional trains.
- Space saving: As metros run underground, they require less space on the surface than conventional trains. This is a major advantage in densely populated areas.
- Reduced congestion and environmental pollution: As an efficient means of mass transportation, the subway helps to relieve road traffic and thus reduce congestion. This also leads to a reduction in air pollution, noise, and greenhouse gas emissions as there are fewer private vehicles on the road.
- Enhanced safety: Subways can be closely monitored and secured, increasing passenger safety and reducing the risk of vandalism or criminal activities.
- Shorter travel distances: Circular connections facilitate continuous journeys without transfers, increasing convenience for commuters. With trains operating in opposite directions in communities on both banks, passengers can reach their destination via the shortest route without transfers (see Fig. 6).

The Lake Area Metro represents a paradigm shift towards modern, resilient transport infrastructure, characterised by efficient, disruption-free operation, and high service frequency. Designed with climate-resilient principles, the metro trains ensure minimal energy consumption and uninterrupted service under all weather conditions. This initiative underscores the commitment to sustainable and resilient urban development, meeting the evolving needs of communities in the Lake Zurich region.

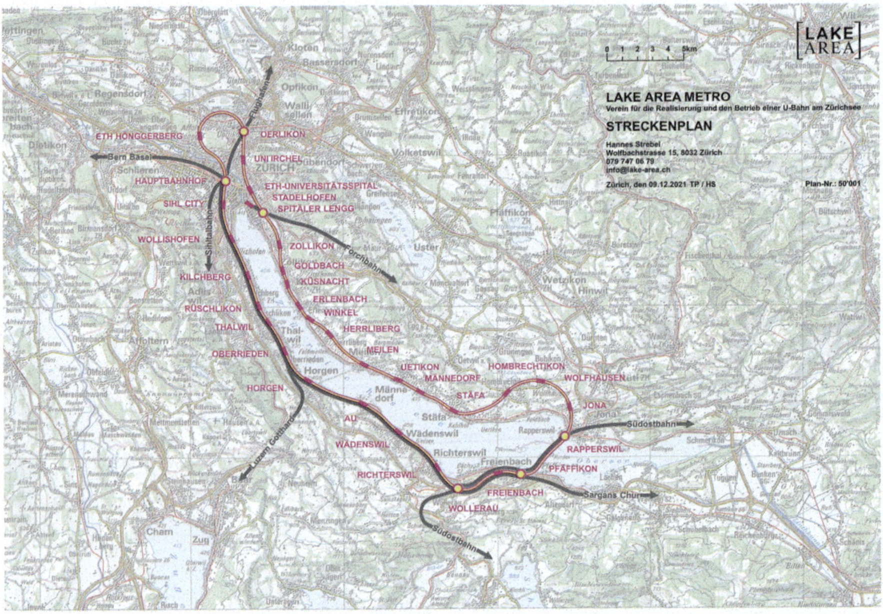

Fig. 5 Route map of the LAKE AREA METRO. The metro (yellow line, purple stations) replaces the overground railroad line on the right bank. The left-bank Zurich—Chur railroad line (black) will be relocated underground between Zürich Hauptbahnhof and Pfäffikon. *Image* Drawing by Hannes Strebel

Fig. 6 The lake is once again an easily accessible recreational area for the local population. *Image* Hannes Strebel

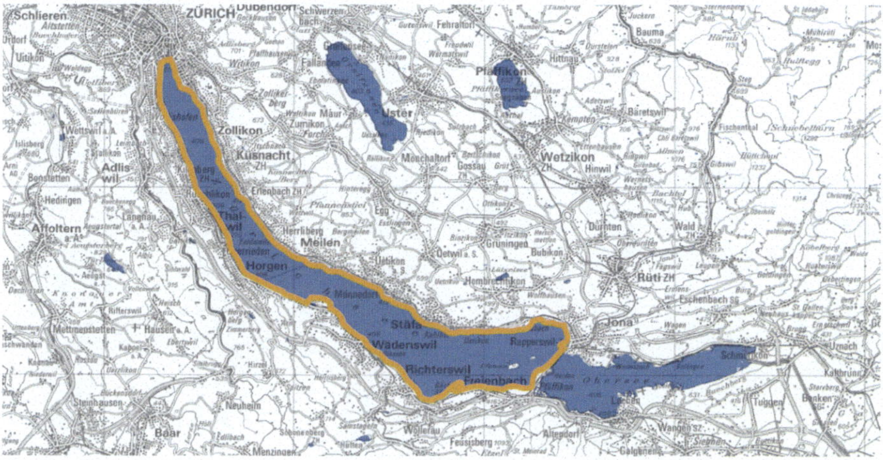

Fig. 7 The existing S-Bahn line will be freed up for a route around the lake for bicycles and e-bikes. *Image* drawing by Hannes Strebel

2.4 Measure 4: The Bike Trail: A Railroad Line for Cyclists Only

Measure 4 introduces the conversion of the existing S-Bahn routes into a dedicated bike trail encircling the lake. With the realisation of the subway ring, the S-Bahn routes will become obsolete and repurposed into a contemporary form: a cycle path reserved exclusively for bicycles and e-bikes.

This initiative creates a health-promoting route for commuting and a recreational pathway for families and avid cyclists alike. By repurposing the former S-Bahn line into a family-friendly cycle path, the region experiences a substantial upgrade in sustainable mobility and recreational opportunities for residents (see Figs. 7 and 8).

The conversion of the former S-Bahn railway line into a family-friendly bike trail represents a unique example of sustainable transformation, aligning with the broader goals of enhancing environmental sustainability and promoting healthy lifestyles within the Lake Zurich region.

2.5 Measure 5: Speed Ferries Across the Lake

Measure 5 introduces the implementation of speed ferries to facilitate efficient transportation across Lake Zurich. Presently, there is a lack of an efficient shipping service between the two shores, with the existing fleet of the Zürcher Schifffahrtsgesellschaft primarily serving tourists.

Fig. 8 A health-promoting work route and a fantastic excursion route for families and sports bikers is being created on the old railroad line. *Image* Hannes Strebel

To address this gap, a fleet of small, manoeuvrable passenger ships will be deployed to offer swift connections across the lake. This concept, which has demonstrated its effectiveness in cities like Venice and Rotterdam for decades, will provide residents and commuters with faster and more frequent cross-lake transportation options.

The introduction of speed ferries will not only enhance connectivity between the shores but also contribute to the overall efficiency and accessibility of transportation within the Lake Zurich region, further improving the quality of life for its residents (see Figs. 9, 10 and 11).

PLAN HECHT stands as a prime example of a comprehensive approach to urban resilience, addressing transportation, environmental, and socio-economic aspects. By redefining the role of transportation infrastructure, it ensures that municipalities are equipped to meet the needs of future generations.

The construction of road tunnels beneath the Pfannenstiel Ridge and Lake Zurich alleviates the traffic burden on municipalities along the right bank while seamlessly integrating them into the national highway network.

The creation of a lakeside path encircling the lake offers residents an unparalleled recreational space amidst a captivating landscape. Efforts to reclaim and enhance shoreline zones contribute to biodiversity conservation and provide residents with valuable recreational opportunities.

The Lake area metro subway ring establishes efficient connections between communities on both shores and key urban centres in Zurich, enhancing accessibility and facilitating sustainable mobility.

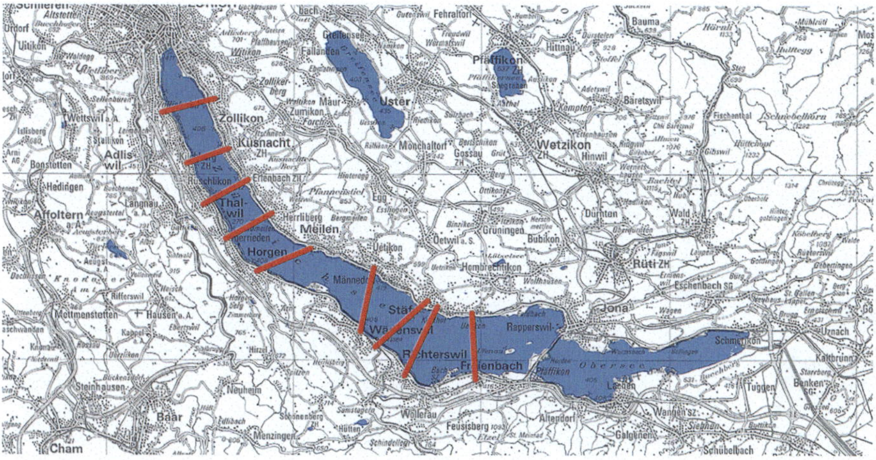

Fig. 9 A fleet of small, manoeuvrable passenger boats is used for the fast connection across the lake. *Image* Drawing by Hannes Strebel

Fig. 10 Cab boats provide a quick connection from shore to shore. *Image* Hannes Strebel

The conversion of disused railroad lines into a health-promoting cycle path represents a unique initiative, promoting active lifestyles and environmentally friendly transportation options.

Additionally, the introduction of fast cab boats further strengthens connections between communities on both shores, fostering closer bonds among residents.

Collectively, these measures bolster the region's resilience to environmental challenges, positioning PLAN HECHT as a model for sustainable urban development that considers the interconnectedness of transportation, environment, and socio-economic factors, ensuring the well-being of present and future generations.

Fig. 11 If you want, you can still glide across the lake on the paddle steamer on warm summer evenings to the sound of riverboat music. *Image* Hannes Strebel

3 Metro Stations as Social Hubs: ETH Zurich Student Competition

Metro stations are not just transportation hubs; they are dynamic social spaces that enhance community life and well-being. Recognising this, the Department of Underground Engineering at ETH Zurich launched a student competition in 2022 for the design of two subway stations: Spitaler Lengg and ETH/Universitatsspital, part of the Lake are metro project.

3.1 Competition Task Definition and Background

The Lake Area Metro Association was established in Zurich on 18 June 2019, to implement and operate a metro system encircling Lake Zurich.

The primary objective is to connect municipalities on both shores of Lake Zurich with each other and with key centres in Zurich, including the health cluster at Lengg and the university campuses at Zentrum and Hönggerberg.

As part of the ETH-based lecture "Planning of Underground Space", the innovative infrastructure project known as the Lake Area Metro was introduced. This project proposes the creation of a metro system encircling Lake Zurich, connecting nearby communities and offering swift access to major centres within Zurich. By relocating traffic underground, the plan aims to generate new open spaces along the lake shore for recreational purposes, thereby significantly enhancing the quality of life for residents in the region. Furthermore, the introduction of the metro line is expected to modernise traffic management, mitigating congestion, heavy traffic, and wait times for residents and other road users.

As part of an interdisciplinary semester project, students specialising in civil engineering, spatial planning, and infrastructure systems collaborated on the design of two underground stations for the LAKE AREA METRO:

1. The ETH-Universitatsspital station. It facilitates access to the ETH University Center campus and the University Hospital. The objective was to seamlessly connect the deep underground Lake Area metro with surface-level streetcar lines.
2. The Spitaler Lenggstation. It provides access to the Lengg hospital cluster, which comprises several healthcare facilities such as the Schulthessklinik, Klinik Hirslanden, Balgrist University Hospital, Children's Hospital, and Zurich University Psychiatric Hospital. The aim is to efficiently link a large volume of passengers to these hospitals via the Lake Area metro, Forchbahn, and streetcar lines.

3.1.1 Spitaler Lenggstation

The city streetcar line 11 and the Forchbahn currently link the city of Zurich and the upper Glatt Valley at Balgrist station with the hospitals in Lengg. However, with the rapid expansion of the hospitals and the increasing number of passengers, these connections are struggling to meet the growing demand. Moreover, there is currently no direct connection to the municipalities on both sides of Lake Zurich (see Fig. 12).

At both stations, there is a suggestion to integrate transportation functions with other uses and approach the topic from spatial planning, architectural, and civil engineering perspectives. Following the evaluation by an expert jury, the top three teams out of the 11 participating teams were recognised and their exceptional concepts and ideas were elaborated upon.

Rank 1—"A Bridge to a Healthy Future"

Student team: Quentin Domézon, Antonella Arnold, Lucian Ramaj, Lars Iten, Elisabeth Brugger

The Spitäler Lengg (formerly Balgrist) ward is situated adjacent to the Lengg hospital cluster, which includes Schulthessklinik, Hirslanden Clinic, Balgrist University Hospital, Children's Hospital and Psychiatric University Hospital Zurich. It is also bordered by the expansive Adlisberg forest and recreational area. At the station, the Forchbahn, linking the upper Glatt valley with the city of Zurich, intersects with the Lake area metro subway line, offering direct connections between lakeside communities, the city, and the hospitals, eliminating the need for passengers to transfer trains.

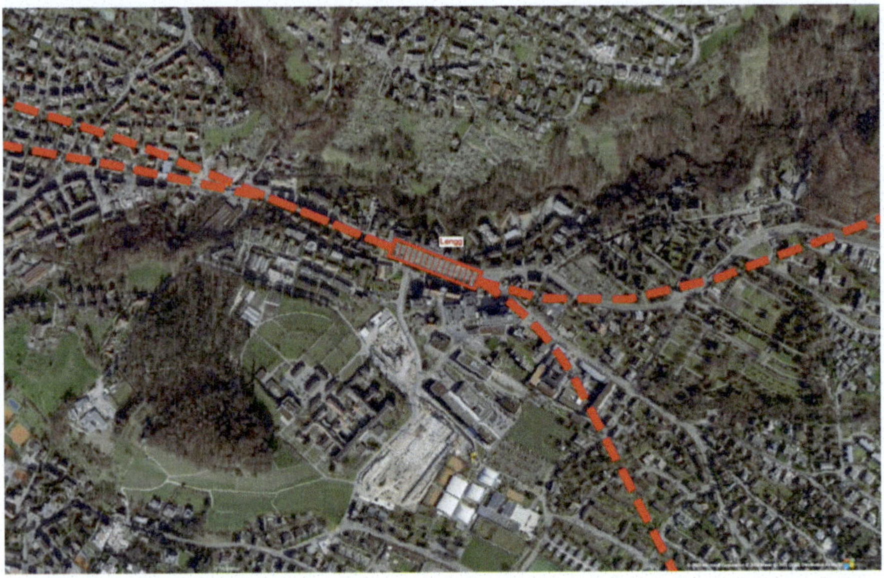

Fig. 12 The picture shows the current Balgrist station. Dashed in red: Tram line 11, Forchbahn and future LAKE AREA METRO. *Image* Student project/copyright: A. Cornaro

Functioning as the central hub of the hospital cluster, the station provides convenient access to the hospitals while facilitating easy entry to the nearby recreational area for hospital users and staff.

The spacious subway area surrounding the station allows for more efficient and cost-effective material supply to the hospitals, as well as accommodating commercial and cultural establishments.

Residents will experience significantly reduced travel times, enabling them to reach the city in half the time previously required (see Figs. 13 and 14).

The concept integrates the hospital buildings in the southwest, connecting them to each other and the central hospital via the subway station. Moreover, it incorporates the valley's topography and the stream, utilising the underground space beneath the hill and establishing a connection to the settlements on the hillside.

The Spitäler Lengg metro station transcends its role as a mere transit area; it functions as a multifaceted hub. It represents the potential for another sustainable site within the broader Lake area metro project.

Rank 2—"Crystal Valley"

Student team: Tim Hess, Lisa Kaufmann, Daniel Alföldi, Sandra Angst, Loris Angst

Students of ETH Zurich were tasked to design a concept for a metro station located within a cluster of hospitals where two underground lines intersect. One key feature

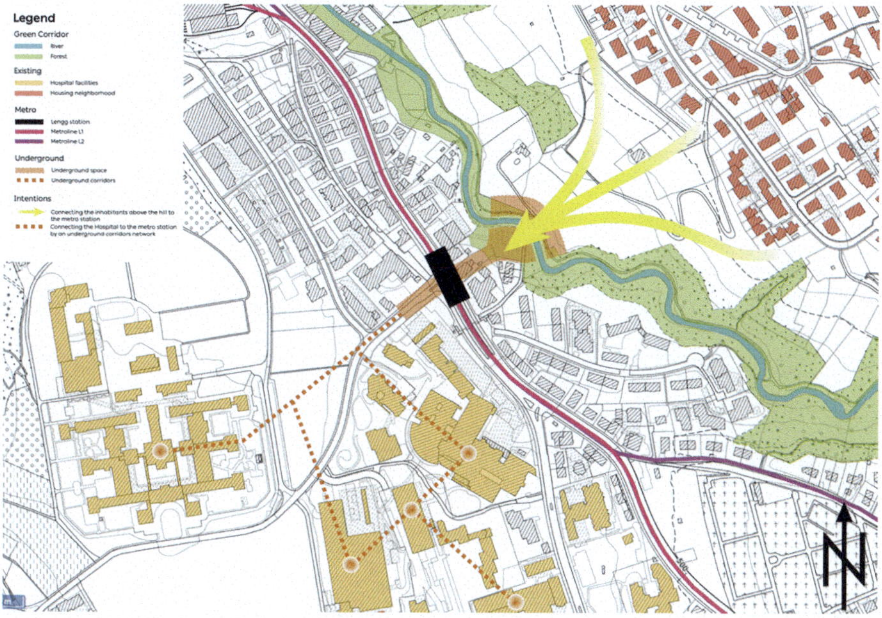

Fig. 13 Situation: The SPITÄLER LENGG station connects the hospitals (dashed red) with the hillside settlement and the Adlisberg recreation area (yellow). *Image* Student project/copyright: A. Cornaro

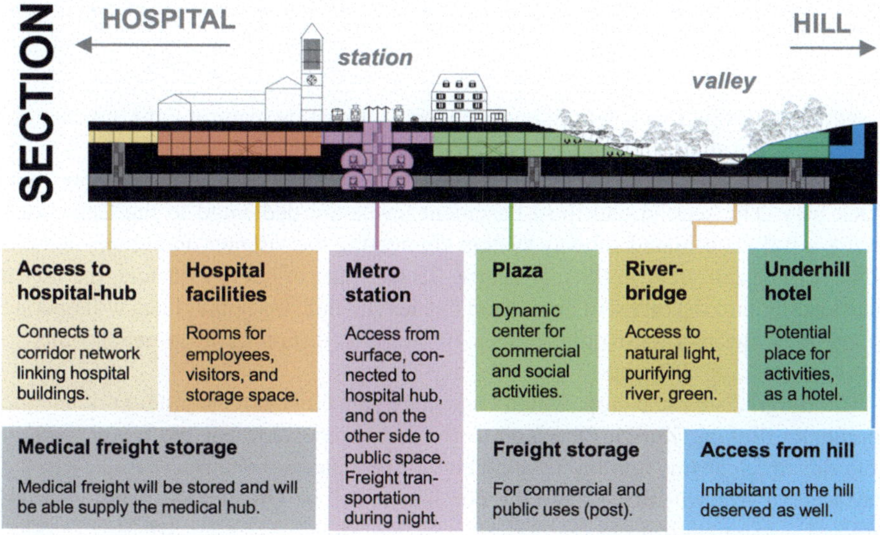

Fig. 14 Cross-section: New services such as a pharmacy, post office, restaurants and leisure facilities can be integrated into the underground spaces. A hotel is being built on the mountainside, by the stream. *Image* Student project/copyright: A. Cornaro

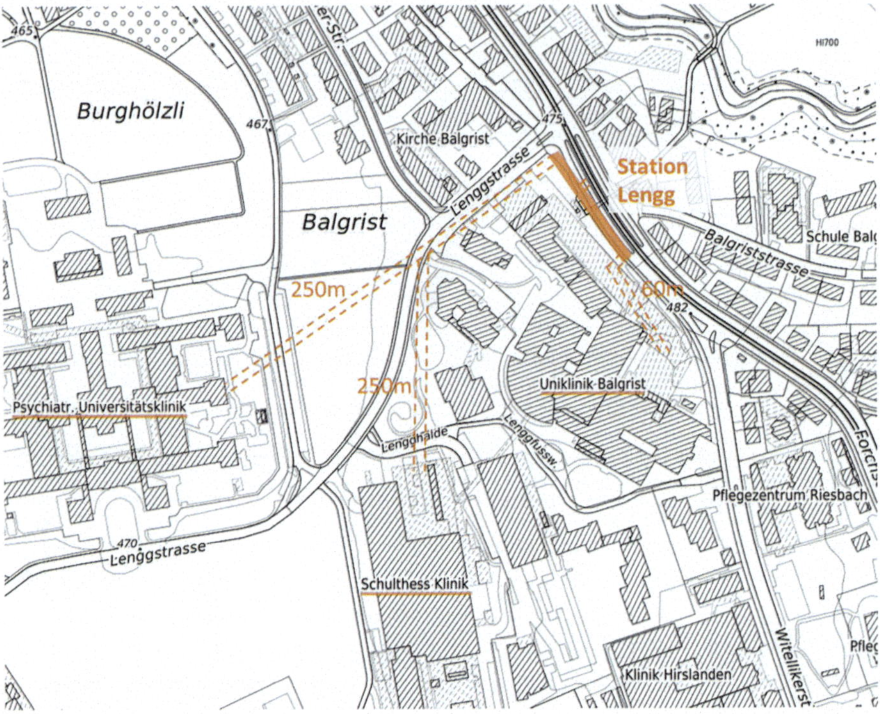

Fig. 15 Situation: The surrounding clinics are accessed underground from the ward. *Image* Student project/copyright: A. Cornaro

sought from design teams was that all concepts must allow natural light to penetrate all levels of the station.

The station consists of three levels. The first basement level houses stores and snack bars. The second and third basement levels are dedicated to the Lake Area Metro, a fully automated subway system connecting the villages and towns on both sides of Lake Zurich with the city centre. To facilitate efficient transfers, the metro lines leading into the city will operate on one level, while outbound lines will operate on another. The two metro platforms will be stacked directly above each other to minimise walking distances.

The Balgrist streetcar stop, which connects to the city's streetcar network and the Forchbahn, linking Zurich Oberland to the city, is located near the metro station. Both the streetcar line and the Forchbahn will remain on the surface. Access to the platforms of streetcar line 11 and the Forchbahn will be via an underpass, ensuring safe and convenient transfers.

A light tunnel situated in the centre of the underground structure will illuminate the shopping and recreation areas on the first basement level, while also allowing natural light to reach the lower levels (see Figs. 15 and 16).

Across the three levels, the station will be used as follows:

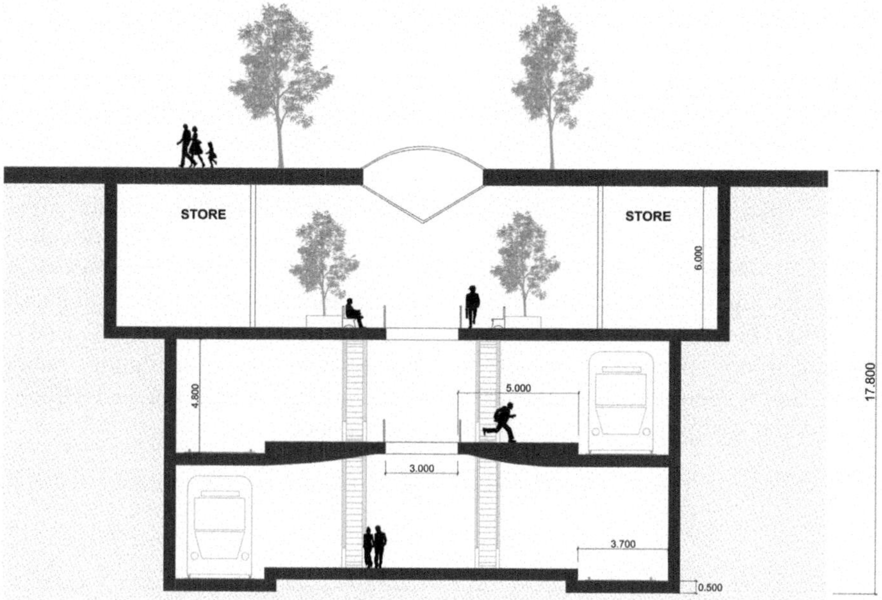

Fig. 16 Section: A shaft brings light down to the lower levels. *Image* Student project/copyright: A. Cornaro

- First basement level: Retail, dining, and entertainment.
- Second basement level: Forchbahn.
- Third basement level: Lake area metro.

The concentration of several hospitals and clinics (Balgrist University Hospital, Schulthess Clinic, Children's Hospital, Hirslanden Clinic, University Psychiatric Clinic) near the new metro station indicates a high expected volume of commuters. Due to the convergence of the Lake Area Metro, Forchbahn, and streetcar, the Lengg station will quickly develop into a highly frequented centre, presenting opportunities for additional uses to be integrated into the station.

The light tunnel fosters connectivity between the different levels and ensures a bright and inviting boarding experience, even on the lowest level. One of the four entrances will take passengers to the lounge area on the top level, providing access to stores, ticket offices, and public facilities. Escalators will offer convenient access to all platforms.

The new subway station will be adjacent to the existing Balgrist streetcar stop. Both Streetcar line L11 and Forchbahn lines share the route above the planned subway station, following the same transport corridor from Stadelhofen to Rehalp. The Forchbahn continues to Esslingen, 15 kms southeast of Zurich. Streetcar line 11 runs between Zurich, Rehalp, and Zurich Auzelg via the main railway station and Oerlikon station.

To facilitate quick transfers between the lines, inbound lines will operate on one level and outbound lines on another. The streetcar line will remain above ground. The two subway platforms will be stacked to minimise distances. Access to the platforms for streetcar lines 11 and Forchbahn will be via an underpass, ensuring safe and convenient transfers. Additionally, this arrangement will improve traffic flow on the busy Forchstrasse by reducing pedestrian crossings.

The Lengg Hospital station is characterised by its surrounding clinics, which serve as significant employment centres and generate commuter traffic, as well as traffic from outpatients and visitors. Direct underground access to the entrances of the three large clinics—Balgrist University Hospital, Schulthess Clinic, and Psychiatric University Hospital—will offer enhanced convenience for all passengers. Underground entrances positioned outside the buildings will ensure continuous public access. The underground accesses to the Riesbach Nursing Center, Hirslanden Clinic, and Children's Hospital must also be taken into account.

Rank 3—"Station Lengg"

Student team: Michèle Arlati, Liliana Cedro, Basil Kretz, Julian Lutz, Florian Münster, Christophe Schlesser

In this design concept, the student team aimed to efficiently connect the hospitals in the health cluster with the city centre, the Zurich Oberland, and the lakeside communities.

To achieve this connection, the team proposed a modern underground stop with the option of transferring to the Forchbahn (see Figs. 17 and 18).

To maximise the limited surface space, all transportation facilities for the metro station are strategically located underground. The entrances and exits, encased in a green steel structure, allow light to filter through, creating a harmonious blend with the surrounding environment. This design ensures that the station not only serves a functional purpose but also integrates seamlessly with its landscape.

The first level, known as the distribution level, hosts stores that cater to the daily needs of passengers and provides space for small markets and cultural events, fostering a lively and vibrant atmosphere.

The Forchbahn operates on the first basement level, connecting Zurich Oberland with the Lengg hospitals and the city, following the same route as the city's streetcar network before descending underground at Lengg station and turning at Stadelhofen station.

On the second underground level, the LAKE AREA METRO links the villages on lower Lake Zurich with the urban centres and the Lengg healthcare cluster, ensuring comprehensive connectivity.

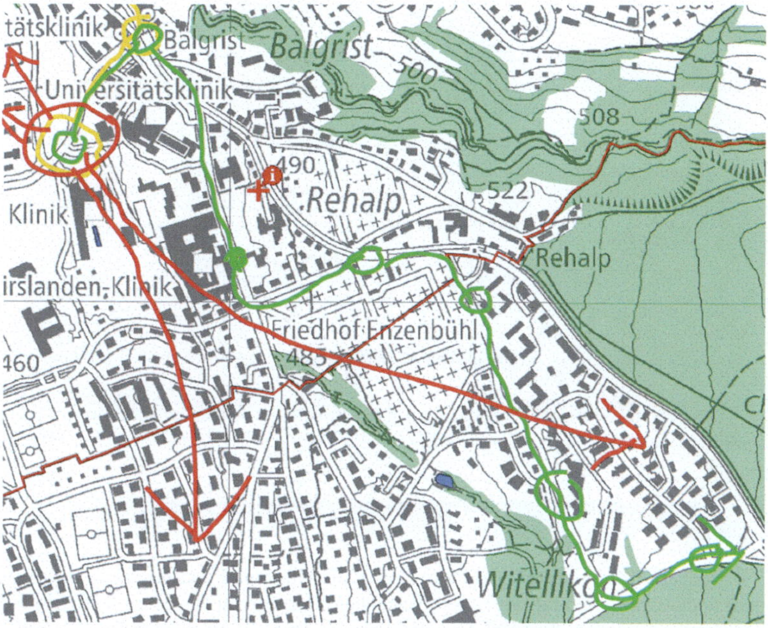

Fig. 17 The Lengg station is the intersection of the Forchbahn feeder line from the Zurich Oberland and the Lake Area Metro, which connects the towns and villages on Lake Zurich and the city centres with the Lengg health cluster. Red line indicates the future Lake Area Metro and Forchbahn lines. From there, it is only a few steps to the Balgrist streetcar station. Green lines indicate private transport routes; yellow indicates footpath between Lengg metro station and Balgrist streetcar station. *Image* Student project/copyright: A. Cornaro

3.1.2 Metrostation ETH-Universitatsspital

The district is bustling with activity, thanks to the plethora of buildings in the area, including university hospitals, which draw in a significant number of employees, students, and patients. Furthermore, the area is densely packed with buildings and infrastructure, adding to the complexity of planning a new subway line and station. The substantial variation in elevation between the routes of the Lake Area Metro and the streetcar lines above presents an additional challenge, requiring thoughtful consideration and innovative solutions to ensure seamless integration and functionality (see Fig. 19).

Rank 1: "The Light Pyramid"

Student team: Lynn Bollinger, Chen Yue, Ramona Hasler, Gabriele Sabato

Fig. 18 Section of design concept: The top level is above ground and covered. Stores for daily needs, small markets and cultural events are located here. The Forchbahn runs on the first basement level as a feeder line from the Oberland. The second basement level is reserved for the LAKE AREA METRO. *Image* Student project/copyright: A. Cornaro

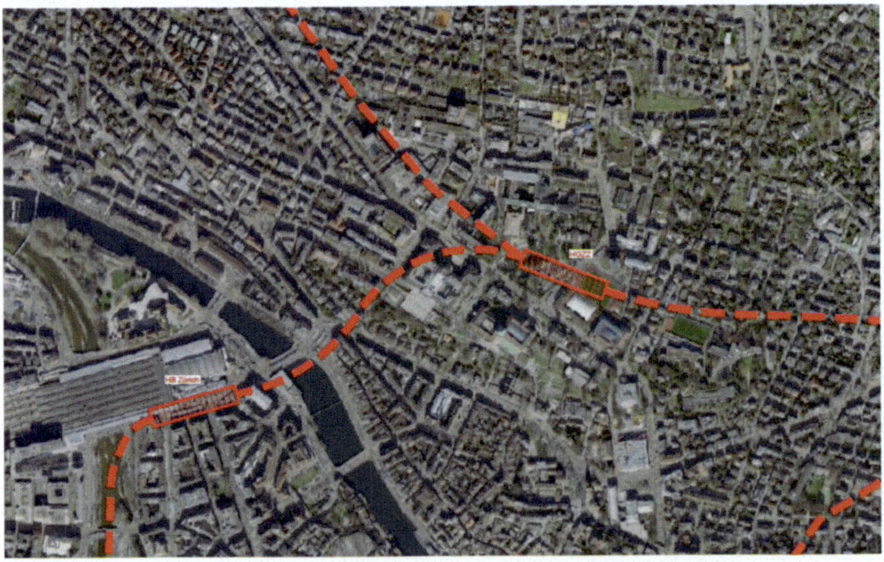

Fig. 19 The future location of the ETH-Universitatsspital station with the Lake Area Metro lines. *Image* Student project/copyright: A. Cornaro

Fig. 20 View from the entrance to the metro station on Gloriastrasse towards the university and model of the central quarter after the HGZZ project/copyright: A. Cornaro

In the heart of Zurich lies the university area, a quintessential quarter defined by its historic buildings like the University Hospital Zurich (USZ), the Swiss Federal Institute of Technology (ETH) and the University of Zurich (UZH).

Amidst this vibrant landscape, the Lake Area Metro project aims to forge a vital link between this central hub and the commuter line, a vision complemented by the concurrent HGZZ project (Zurich University Center).

The HGZZ project, currently in the conceptual planning phase, envisions a revitalisation of the neighbourhood, centred around a park adjacent to Gloriapark Hospital. This collaborative effort, involving USZ, ETH, and UZH, aligns seamlessly with the proposed "Light Pyramid" project, harmonising with the area's sustainable redevelopment (see Fig. 20).

Positioned at the southeast corner of the envisioned park, the subway station's design prioritises safeguarding above-ground spaces, notably the new Gloriapark and key thoroughfares. The innovative concept of an inverted pyramid, rotated 180°, yields a striking architectural feature—a vast subterranean space, flanked by concentric terraces, seamlessly linking surrounding buildings via four corridors (see Fig. 21).

This meticulously planned underground complex is divided into four levels, with the pinnacle adorned in a striking hue of red, offering recreational amenities amidst lush greenery, fostering social interaction and community engagement. Descending further, the lower levels house the functional core of the subway station, complete with lateral platforms facilitating swift passage to the surface via state-of-the-art elevators. Ascending from the depths, visitors emerge into a sunlit courtyard, bathed in natural light and framed by verdant foliage, offering pathways to the area's prominent institutions and direct access to Gloriapark above (see Fig. 22).

Spanning approximately 90 m beneath the earth's surface, this subterranean marvel is designed with a keen focus on accessibility, leveraging the inverted pyramid's form to optimise the ingress of natural light. Strategically situated at a pivotal transportation nexus, the station's location ensures seamless integration with existing streetcar lines, promising a transformative transit experience that bridges Zurich's rich past with its promising future.

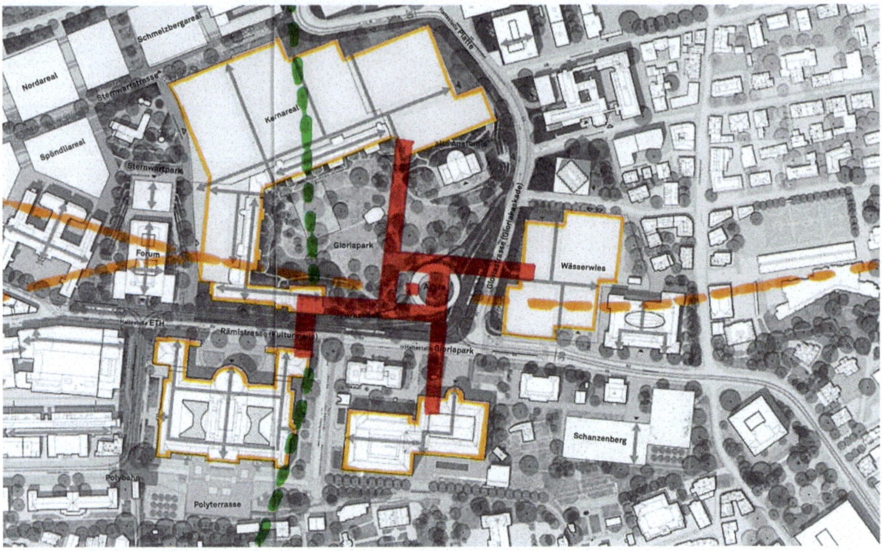

Fig. 21 The redesigned Zurich University area includes a proposal for an underground metro station. Orange: the new metro line; green: the existing SBB Weinberg tunnel; yellow: the existing and new buildings of USZ, UZH and ETH. Red: the metro station with the four corridors for underground access to the surrounding buildings. *Image* Student project/copyright: A. Cornaro

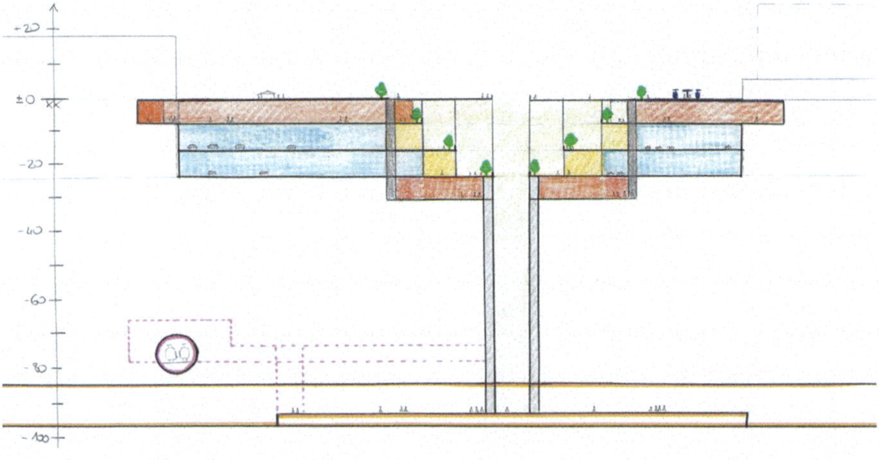

Fig. 22 Sketch of the "inverted pyramid" concept for the new underground structure. Parking spaces (blue) and bars or rest areas for people are planned on the second and third levels. *Image* Student project/copyright: A. Cornaro

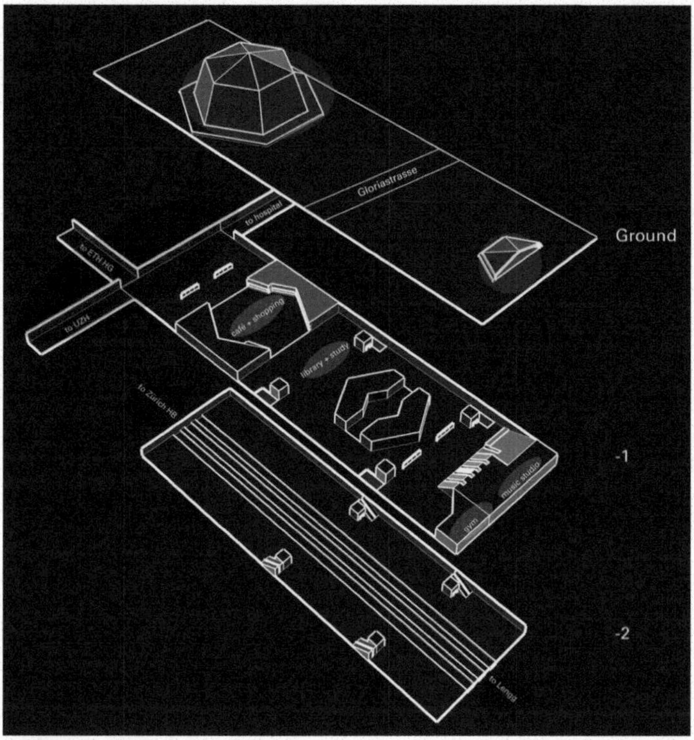

Fig. 23 The station is divided into three levels. The metro trains operate on the lower level. The middle level is open for various uses and will be a meeting point for all people who frequent the station. *Image* Student project/copyright: A. Cornaro

Rank 2: "Meeting Hub"

Student team: Anna Hodel, Simon Käslin, Kim Bottinelli, Wenhui Wu, Timon Stettler, Yanick Graf

The concept for the HGZZ station was based on the fusion of multifaceted functionality. Beyond merely facilitating access to the bustling university district and seamlessly integrating it into Zurich's urban fabric, the station aspires to be a versatile hub catering to diverse needs. Designed to transcend mere transit, this space is envisioned to be a vibrant nexus serving the vast student population as well as the wider public. It is conceived as a dynamic gathering spot, fostering interaction and dialogue among its diverse denizens—a place where ideas converge, connections are forged, and community thrives. In essence, the HGZZ station emerges not just as a conduit of mobility, but as a catalyst for social engagement and collective enrichment (see Figs. 23, 24, 25 and 26).

Given the constraints of limited space for expanding the university campus in this bustling district, an underground extension emerges as the clear and practical

Fig. 24 The top level is at the same height as the existing park. It provides the floor below with daylight. *Image* Student project/copyright: A. Cornaro

Fig. 25 The middle level offers space for public libraries, ETH and UZH study rooms and university sports facilities. *Image* Student project/copyright: A. Cornaro

Fig. 26 The lowest level is used exclusively for access through the LAKE AREA METRO. *Image* Student project/copyright: A. Cornaro

solution. With an eye towards optimising the station's functionality, meticulous plans include direct pedestrian connections to key institutions such as ETH, the university, and the university hospital. These underground passageways not only enhance accessibility but also seamlessly integrate the station into the fabric of the surrounding academic and healthcare landscape. By fostering direct links between these vital nodes, the station serves as a linchpin for facilitating smooth and efficient movement within the vibrant university district, enriching the experience of students, faculty, and visitors alike.

Rank 3: "Station HGZZ"

Student team: Vanessa Reiser, Ian Chamberlain, Muyang Jiang, Pascal Hodel

Positioned beneath the prestigious institutions of ETH Zurich and the University of Zurich, the station serves as a vital artery, seamlessly linking these academic powerhouses to communities along the shores of Lake Zurich and key urban centres like the main railway station, Irchel University, and the Lengg Health cluster via the Lake Area Metro (see Fig. 27).

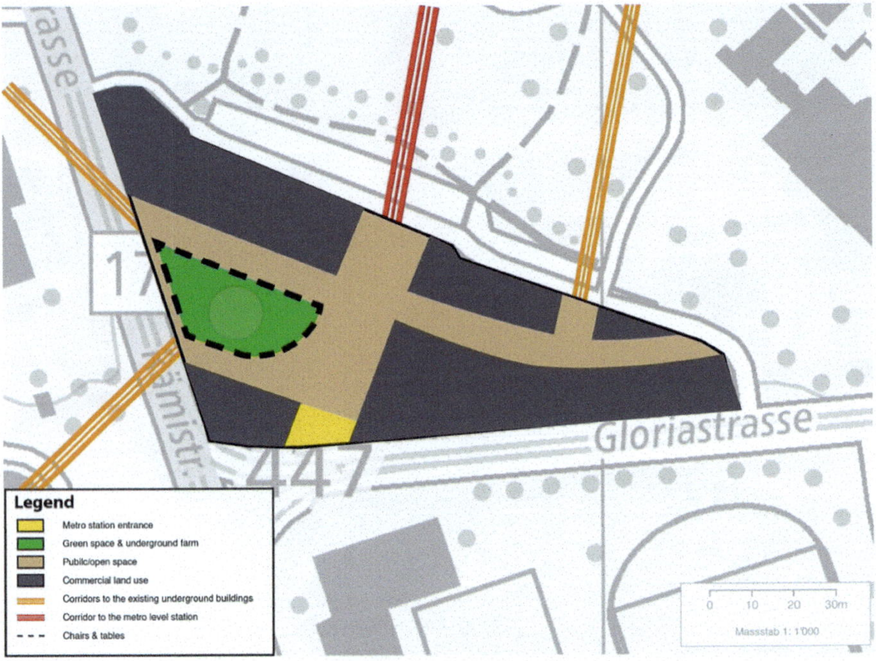

Fig. 27 The station is situated at the intersection of Rämistrasse and Gloriastrasse in the south-eastern corner of Gloria Park, *Image* Student project/copyright: A. Cornaro

Comprising two levels, the station offers a dynamic fusion of convenience and leisure. The upper level, a vibrant mixed-use public space, teems with shops, eateries, and verdant greenery, inviting visitors to linger and engage. Below, the lower level is dedicated exclusively to Lake Area Metro subway transportation, accessible via four escalators and a staircase from the upper level.

The most significant buildings—University Hospital Zurich, ETH Zurich, and the University of Zurich—will be seamlessly connected to the station through pedestrian tunnels, ensuring efficient and uninterrupted mobility within the academic precinct.

Notably, alterations to the original plan were necessitated by congestion beneath Zurich Main Station, prompting a strategic deviation in the route. Consequently, the Lake Area Metro tracks now traverse north–south at a 90-degree angle to the existing tracks, mitigating challenges posed by crossing the intricate network of tunnels beneath the Limmat River to the east. This adaptive approach ensures efficient connectivity and smooth passage, underscoring the station's pivotal role in Zurich's evolving urban landscape (see Fig. 28).

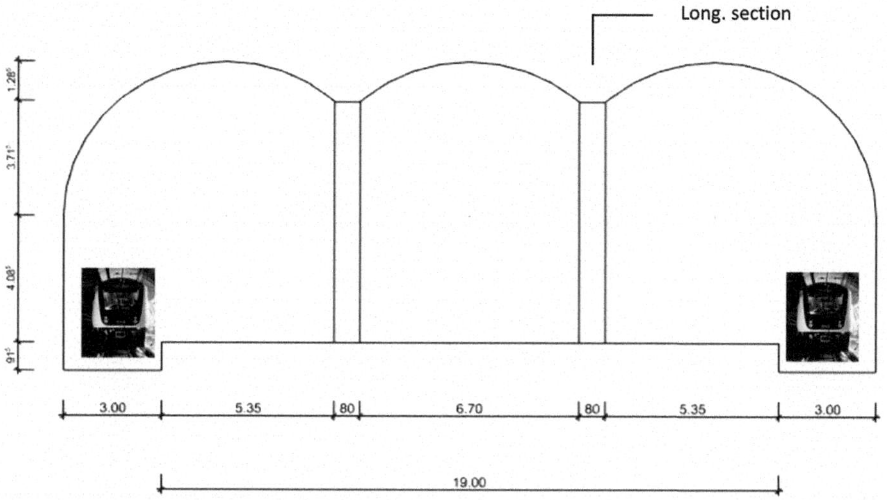

Fig. 28 Cross-section with the dimensions of the subway level, *Image* Student project/copyright: Antonia Cornaro

4 Discussion

The Lake Area Metro represents a paradigm shift towards a modern, resilient transportation infrastructure. Designed for efficient, disruption-free operation with high-frequency service, the metro incorporates climate-resilient principles, ensuring minimal energy consumption and resistance to extreme weather conditions. This ensures uninterrupted service regardless of climatic challenges.

Additionally, repurposing former suburban railroad lines as family-friendly cycle paths underscores the project's commitment to sustainable mobility and recreational opportunities for residents. This transformation supports a shift away from car dependency, promoting healthier, more active lifestyles.

Beyond transportation, metro stations are envisioned as vibrant hubs for social interaction and commerce, enriching the urban fabric and enhancing community well-being. By reclaiming above-ground space for ecological restoration, the project promotes biodiversity and green urban infrastructure, further strengthening the region's resilience to environmental challenges. The Lake Area Metro exemplifies a holistic approach to urban resilience, addressing transportation, environmental, and socioeconomic dimensions to create vibrant, livable communities for present and future generations.

The introduction of unconventional ideas, particularly by students of ETH Zurich, often triggers a defensive attitude, and the Lake Area Metro project was no exception. Critics argued it was unnecessary, lamented the loss of scenic train rides, and doubted its affordability. However, careful consideration of the proposals soon revealed not only the advantages of faster connections and better accessibility but also the potential profitability, which piqued investor interest.

Stakeholders quickly recognised the unprecedented opportunity for optimal access to the Im Lengg hospital hub provided by the metro. It is hoped that political bodies will draw their conclusions and set the course for early realisation. Without decisive action, the region risks worsening congestion in existing transport modes, leading to their eventual collapse and serious consequences for the population, the economy, and industry.

In essence, the Lake Area Metro's innovative approach and forward-thinking solutions present a compelling case for immediate implementation, ensuring sustainable growth and enhanced quality of life for all residents.

Creative Depths: Art, Innovation, and Reimagining the Underground

Rock Skin—A Process Toward Ecological Empathy

Meng Li and Linda Zhang

Abstract This chapter explores the concept of ecological empathy through the case study of the Rock Skin pavilion, an installation by Studio Pararaum exhibited as part of the 2021 Seoul Biennale of Architecture and Urbanism. Rock Skin responds to the disconnection between the built environment and its geological origins, with the aim of reconnecting city dwellers to the site of underground material extraction. Central to the project is the application of the Connectedness to Nature Scale (CNS) in the context of architecture. By integrating experiential and psychological sustainability into architectural design, the project seeks to cultivate a deeper bond between inhabitants and the natural world, ultimately fostering sustainable behaviors and environmental stewardship in the day-to-day lives of inhabitants. Using mixed methods, including quantitative surveys and qualitative interviews, the study evaluates the impact of the Rock Skin installation and its design processes in enhancing environmental connectedness for promoting eco-conscious practices. The research explores the role of participatory design in fostering ecological empathy, positing that involvement in the design process correlates strongly with increased ecological empathy. It evaluates and recommends amplifying the role of participatory design in architectural education and practice.

Both authors contributed equally to this article.

M. Li (✉)
Department of Architecture, Swiss Federal Institute of Technology Zurich,
Wolfgang-Pauli-Strasse 5, 8049 Zurich, Switzerland
e-mail: meng.li@arch.ethz.ch

L. Zhang (✉)
School of Architecture, University of Waterloo, 7 Melville St, Cambridge, ON N1S 2H4, Canada
e-mail: linda.zhang@uwaterloo.ca

© The Author(s), under exclusive license to Springer Nature Singapore Pte Ltd. 2025
M. Agrawal (ed.), *Underground Spaces for Climate Resilience and Sustainability*,
Advances in 21st Century Human Settlements,
https://doi.org/10.1007/978-981-96-0547-7_13

1 Introduction

This chapter introduces the concept of ecological empathy and situates its potential to shape the future of city-building through the case study of Rock Skin, pavilion installation by Studio Pararaum, exhibited at the 2021 Seoul Biennale of Architecture and Urbanism. Rock Skin is a novel approach toward the building practice and design of urban living spaces that responses to the long-standing detachment between building and sourcing. This method introduces an innovative intermediary—the rock skin—which bridges 'a fragment of the mountain' and 'a building block.' By doing so, it aims to establish a new relationship between urban material consumption and natural mountain resources. Much like capturing a photograph, this technique involves preserving the mountain's surface 'in time' using thin layers of natural latex [1].

As cities expand and infrastructure needs amplify, the extraction of natural resources progresses at ever-accelerating paces. City dwellers, however, remain largely unconcerned in their daily routines about the toll that these extractive processes have on the Earth's fragile ecosystems. Mountains are excavated and even moved, rivers rerouted and even reversed, and landscapes irrevocably altered in the relentless pursuit of urban expansion and capitalist extraction. Yet, amidst this transformation, the geological essence of these landscapes remains largely invisible, obscured by neoliberalist pursuits for so-called progress and efficiency. It lies hidden behind the veneer of concrete and steel that defines the modern cityscape, concealed underground—out of sight and out of mind.

In response, Rock Skin reframes the relationship between cities and their geological origins. By reconnecting city dwellers with the geological origins of their built environment, Rock Skin explores how ecological empathy—a deeper appreciation for the interconnectedness of all living things—can challenge both prevailing notions of resource extraction and human behavioral responses to resource consumption.

Ecological empathy embodies a profound interconnectedness between human actions, architectural interventions, and the natural environment. It transcends mere awareness of the climate crisis, toward a deeper relational understanding of responsibility to care for the ecological material landscape upon which our cities are built and our mutual well-being depends. By integrating experiential and psychological sustainability into architectural design, the Rock Skin project aims to nurture a deeper connection between inhabitants and the natural world, fostering sustainable behaviors and environmental stewardship. Positioning the underground, mined, and subterranean geologic landscapes as architectural heritage, Rock Skin offers a holistic approach to city-building that invites us to reconsider our relationship with the natural world through participatory design action, affect, and material processes (see Figs. 1 and 2).

Fig. 1 The making of Rock Skin inside a historical underground mine in Schleitheim, Switzerland. Photograph by Stefan Müller, 2021. Courtesy of Studio Pararaum, 2021

Fig. 2 Physical 1:50 scale model of the 'Rock Skin Pavilion,' designed to reconnect urban residents with the geological underground. This installation was featured at the Seoul Biennale of Architecture and Urbanism (SBAU) in 2021. Photograph courtesy of Studio Pararaum, 2021

2 Sustainability: Understanding the Gap Between Attitude and Behavior

Despite growing awareness of environmental issues, many individuals struggle to translate their beliefs into tangible actions that promote sustainability [2]. Aligning behaviors to attitudes and beliefs requires more than just innovative design solutions—it necessitates a fundamental re-evaluation of how we measure and monitor the impact of architecture on human behavior and attitude. Traditional metrics in architecture for sustainable design, such as green building rating systems, often focus on quantitative criteria related to building materials and construction processes, overlooking the crucial role of human psychology and behavior in shaping a building's environmental performance post-occupancy, including the actions of its inhabitants. As such, an interdisciplinary and holistic approach is needed to develop new metrics and evaluation tools that capture the complex interplay between design, human behavior, and environmental outcomes.

2.1 Connectedness to Nature Scale (CNS)

The dissonance between human attitudes, values, and their resulting behavior is a widely studied area in psychosocial research, however, it remains underexamined in the fields of architecture and design. One key framework for evaluating ecological empathy is the Connectedness to Nature Scale (CNS) [3]. Developed within the field of environmental psychology, the CNS measures an individual's psychological connection to the natural world (environmental connectedness) and correlates with cognitive behavioral changes that support sustainable practices.

Rock Skin leverages the connectedness to nature scale (CNS) in an architectural context with the hypothesis that effective ecological behavioral change requires architecture to not only be factually sustainable (as set out by green rating systems) but also be experientially, phenomenologically, and psychologically sustainable—connecting inhabitants to nature and building ecological empathy through lived experience. By integrating evidence-based mixed methods assessments, the Rock Skin project evaluated the impact of a design's connectedness to nature and occupants' willingness to adopt sustainable practices.

Through the CNS, this research seeks to expand the definition of sustainable architecture to include psychosocial and behavioral dimensions. Drawing links between environmental psychology, architectural phenomenology, and architectural heritage, Rock Skin examines the tangible impact of architectural design on day-to-day public behavior, socio-political vision, and public imagination.

3 Methodology

We used a mixed methods research approach focusing on quantitative surveys and qualitative interviews. Through empirical surveys and interviews with study occupants and participants, the project evaluated the effectiveness of the Rock Skin design process in promoting sustainable behavior and environmental awareness.

Our surveys evaluated the CNS in three distinct design scenarios to understand their effectiveness in promoting sustainable behavior (Fig. 3). This included (1) a baseline urban environment, (2) a baseline natural environment, and (3) the Rock Skin Pavilion design. The two baseline environments represent contrasting settings. The (1) typical interior environment reflects the built environment prevalent in urban areas, characterized by artificial lighting, air conditioning, uniform smooth surfaces, and limited access to natural elements. In contrast, the (2) natural environment represents a setting immersed in nature, featuring elements such as contrasting light, highly textured surfaces, and natural landscapes. By evaluating these two environments, we compare and situate how individuals respond to the Rock Skin design with varying levels of natural elements and ecological cues. Through the Rock Skin prototype, we established a conceptual framework to better understand the correlation between the built environment, connectedness to nature, and sustainable behavior.

Next, we examined the impact of participatory design: what impact does becoming involved in the design, material sourcing, and building of the architectural setting have on an inhabitants' ecological empathy and day-to-day actions toward environmental stewardship? Building on existing literature in design practice and psychology [4], we hypothesize that involvement in the design process is strongly correlated to ecological empathy—as measured by the CNS. If this holds true, the architectural design process itself becomes an important site for building ecological empathy that can serve as a powerful catalyst for sustainable behavior change. This would also help better understand the role of citizen participation [5] in the design and construction of our built environment.

We first studied the impact of participatory design with a group of Interior Design students at the Toronto Metropolitan University to evaluate the impact and role (if any) of being involved in the process of design on CNS scale. Through participatory design and fabrication workshops over a semester-long design studio taught by Prof. Linda Zhang, students applied the Rock-Skin design process to make their own interior designs, including material studies, material sourcing, physical prototyping, and model making. We evaluated their CNS before and after the class with the hypothesis that it would increase their ecological empathy. The level of increase from participatory design was then compared to the level of increase in CNS (if any) of the Rock Skin environment—to assess the impact of participatory involvement in the design process when compared to traditional non-participatory design. We then conducted additional qualitative and quantitative analysis of the CNS increase of the design team of the Rock Skin Pavilion at SBAU (see Fig. 4).

Fig. 3 360-degree image of each of the three environments. Top: (1) a baseline urban environment, 360 scan by Studio Pararaum, 2021; Middle: (2) a baseline natural environment, 360 scan by Studio Pararaum, 2021; Bottom: (3) the Rock-Skin Pavilion design, 360 scan by Subin Kim design assistant of Studio Pararaum and team lead for installation of Rock Skin for SBAU, 2021

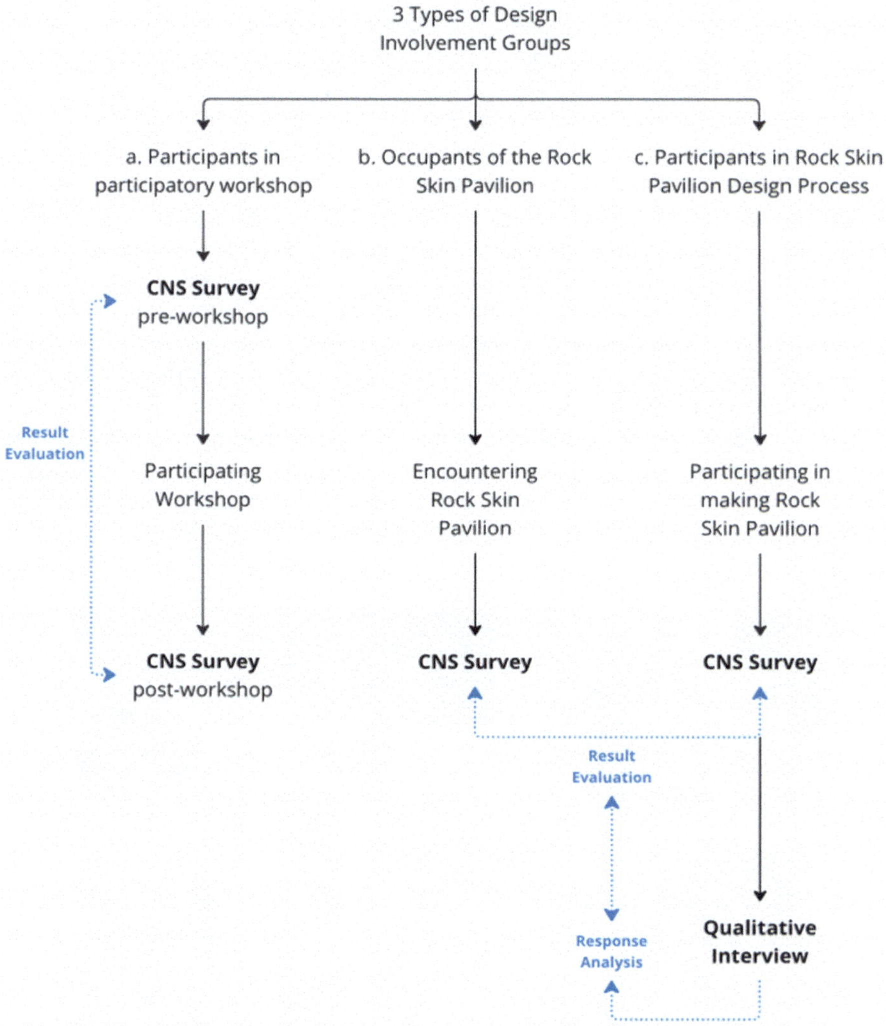

Fig. 4 Diagram showing evaluation and analysis of CNS on groups with different types of involvement

4 Rock Skin: Connecting to Nature Through Material Stories

The Rock Skin design process works closely with phenomenological and indexical qualities of materials—the way matter communicates without words, i.e., via material stories. While sustainable design has advanced significantly in technological dimensions, such as developing low-carbon materials, the affective aspects of material development remain under-explored. Yet, the affective dimension [6]—how materials can evoke emotions through bodily experiences—offers the best opportunity for building ecological empathy [7].

Materials are the very artifacts and markers of time, distance, and place. They travel from their sites of resource to the cities they build. As people hold and touch the matter of the earth that surrounds us, such daily environments have great potential to mitigate the negative impacts of climate crisis via human behavioral changes—both implicit and explicit. Standardized green building materials may have low embodied carbon and are defined as sustainable, however, they are often still generic in their appearance, completely detached from their natural material story. Many of these green products take on the same aesthetics as their traditional counterparts: outside of their certification labels, a viewer would not be able to tell the difference between them (see Figs. 5 and 6).

The Rock Skin project makes use of this lost opportunity by reconnecting the physical appearance and tactile qualities of materials with their geological sources, which have been shaped over millennia. The approach aims to foster a deeper connection between the inhabitants and the natural environment through the very materials

Fig. 5 Collection of images on material approaches toward connectedness between geological context and built environment, serving as inspiration for conceptual framework development of Rock Skin. Collage by Studio Pararaum, 2020

Fig. 6 A fictional model narrating an encounter with geologic mountains through architecture. Photography by Studio Pararaum, 2020

themselves that are sourced to build our cities. To do so, the material process of Rock Skin makes visible the geologic mountains without further depletion of the natural resources. Rather than directly extracting natural stones from the quarry, a new agent is introduced, mediating between 'a fragment of the mountain' and 'a building block.' First, thin layers of the natural latex are applied to the rugged surfaces of geological formation. Once cured, the latex mold is gently peeled off, leaving the mountains virtually 'untouched.' Embedded with the geological texture, this flexible latex mold is what Studio Pararaum termed the Rock-Skin-Mold (Fig. 7).

These natural latex Rock-Skin-Molds serve as a medium for the transfer of geological memory and stories. They act as molds to cast bioplastic (a non-petrochemical-based plastic made from waste byproducts), resulting in thin, translucent Rock-Skin-Casts (Fig. 8). Thus, the Rock Skin design process simultaneously minimizes the impact of resource extraction while maximizing the connection to geological memories and site. As flexible frameworks for casting, Rock-Skin-Mold can be reused up to 100 times.

In turn, thin layers of liquid bioplastic are applied onto the latex mold. Once solidified, they are demolded from the latex mold and the resulting bioplastic casts quite literally embed the intricate texture and haptic qualities of the geological texture. Not pretending to be a rock, these translucent casts appear to be ephemeral, alluding not only to the lightness of its environmental footprint but also to the fragility of our ecological system. Through this process, Rock Skin captures the texture, form, and stories of mountains and their underground landscapes, offering the potential of re-embedding their memory within the fabric of urban architecture.

Fig. 7 Small-scale experiment illustrating the process of making the Rock-Skin-Mold. Photograph by Studio Pararaum, 2020

Fig. 8 Small-scale experiment displaying the resulting Rock-Skin-Cast in bioplastic, created using the Rock-Skin-Mold. Photograph by Studio Pararaum, 2020

5 Seoul Biennale of Architecture and Urbanism: Reconnecting City and the Underground Mine

In response to the call of Seoul Biennale of Architecture and Urbanism 2021, we contextualized the material process of Rock Skin in the historical underground mine, today's Gipsbergwerk museum or gypsum mining museum, in Schleitheim, Switzerland (Fig. 9). The site, renowned for its natural gypsum extraction dating back to the early eighteenth century, saw its first subterranean tunnel established in 1790, marking the transition from surface to underground mining [8]. Within the mining cave, the rugged surfaces reflect the evolving techniques and extensive labor involved in the long history of gypsum mining (Figs. 10 and 11). This raw material has been continuously extracted for the construction of the Rheinau Monastery interior, supporting the growth of surrounding cities until the 1940s. Over centuries, human activities have induced irreversible transformations in geological sediment formed over millions of years, creating countless voids within the mountain—hidden from city dwellers and disconnected from the urban fabric of life.

Using natural latex rubber, the sensual texture of Schleitheim gypsum mine was captured in the reusable molds, (Figs. 12, 13 and 14), with which, eighteen large-scale shells were cast in bioplastic. The Rock-Skin-Casts were then assembled into a canopy and raised by the wooden columns on-site at the Seoul Biennale of Architecture and Urbanism (SBAU), creating a hovering memory of the geologic cave above wandering visitors. Like the photograph, the Rock Skin installation at Seoul transports the natural history of the underground mine through a material snapshot, or imprint, in time. In turn, as we encounter the casts, we can still experience the reality of the underground void through its texture, even though the void isn't there directly. In a sense, the Rock-Skin-Casts are to the underground mine what the trees are to the wind: you need trees to hear the wind. In parallel, the Rock-Skin-Casts help us perceive the existence of the void below (see Figs. 15, 16, 17 and 18).

6 CNS Survey Findings: Toward Ecological Empathy

In the following sections, findings and future recommendations based on the quantitative (CNS survey) and qualitative (interview) results of the study are highlighted. Negatively worded items (questions 4, 12, 14, and optional question 15 indicated in Table 1) were reversed prior to analysis for Likert scale directional alignment. Strongly agree (green) represents the most connected to nature and strongly disagree (yellow) represents the least. When measuring participatory design impact, 14 questions from the standard CNS survey of Stephan Mayer and Cynthia McPherson Frantz were given. When measuring the three design scenarios, a shortened 10-question version of the standard CNS survey was used and two additional architectural questions were given.

Fig. 9 Exterior view of the historical access to the underground gypsum mine, now the Gipsbergwerk museum (Gypsum Mining Museum) in Schleitheim, Switzerland. Photograph by Félix Michaud, 2022. Commissioned by and courtesy of Studio Pararaum

(15) I often felt the natural elements in this environment were artificial and disconnected from nature.

(16) I feel more connected to nature after exploring this environment.

These two questions assess the direct perceived impact of the specific environment. Survey results demonstrate that:

1. **The Rock Skin Design Installation fosters more connection to nature than the generic interior (baseline 1) but less than the natural cave (baseline 2).**

The baseline 2 in a nature setting produced the most connection to nature. However, the CNS results from the Rock Skin Design installation were comparable to those

Fig. 10 Interior view of the access tunnel within the Gipsbergwerk (Gypsum Mining) in Schleitheim, Switzerland. Photograph by Félix Michaud, 2022. Commissioned by and courtesy of Studio Pararaum

of the natural cave baseline setting. Notably, Rock Skin shows the most significant improvement in connectedness to nature in question 16. Compared to both baseline environments, it indicated a higher impact and felt a change in inhabitants (see Fig. 19).

2. **Significant Impact of Participatory Design on Increased Connectedness to Nature**.

The participatory design workshops, conducted from January to April 2021, involved 14 students who met once a week for 6 h over 12 weeks, totaling 72 contact hours. We hypothesized a correlation between design participation and connectedness to nature. Interestingly, the data revealed a remarkably strong increase in students' connectedness to nature, as evidenced by the significant shift in responses on the Likert scale. Students showed a sharp rise in instances of strong agreement, reflecting a notable certainty in their change of feelings. This outcome is particularly striking when compared to the CNS survey responses to the three scenarios mentioned earlier (see Fig. 20).

Similarly, we observed increased CNS scores in the students ($n = 8$) that participated in the material research, design, and prototyping of the Rock Skin Pavilion. These two participatory design case studies indicate a strong affinity between participatory design process and building empathy for nature (see Fig. 21).

Fig. 11 Interior view of the underground mining vault inside the Gipsbergwerk (Gypsum Mining) in Schleitheim, Switzerland. Photograph by Félix Michaud, 2022. Commissioned by and courtesy of Studio Pararaum

7 Qualitative Findings: Behavioral Changes Toward Environmental Stewardship

Our quantitative findings demonstrated that participatory design plays a crucial role in fostering ecological empathy among occupants. By engaging people in the making, designing, and fabricating of the SBAU Rock Skin Pavilion, we examined how the

Fig. 12 Application of natural latex to make the Rock-Skin-Mold in Gipsbergwerk Schleitheim, Switzerland. Photograph by Stefan Müller, 2021. Courtesy of Studio Pararaum

participatory design process enhanced the understanding of material sourcing and lifecycle, thereby promoting sustainable behavior among the student design team.

One of the qualitative findings was participatory design's ability to bridge the gap between abstract sustainability principles and tangible implementation in the day-to-day. By actively involving people in decision-making processes, participatory design creates opportunities for meaningful co-learning and collaboration.

Fig. 13 Removal of the Rock-Skin-Mold in, preserving the surface of the underground mine. Photograph by Stefan Müller, 2021. Courtesy of Studio Pararaum

Prior to the project, I had a hard time visualizing how to make things, or ideate without consuming and extracting irresponsibly. I was able to better understand and reflect on how one can practice, innovate and create in a manner that is in line with my individual values and consider nature as a stakeholder in my projects.

I think working with my hands and having them shape the bioplastic forms via subtle touch or heat or special technique made me feel the most connected to nature. In a way it brought out this child-like sense of self where discovery via hands-on learning is prioritized and encouraged. It was an imperfect process that had its challenges and storms, but resulted in a beautiful end product.

Fig. 14 Large-scale Rock-Skin-Mold demolded, embedding the material existence of the underground mine without depletion. Photograph by Stefan Müller, 2021. Courtesy of Studio Pararaum

Existing literature has also established a positive impact of participatory design on occupants' connection to nature and their adoption of sustainable practices. For example, projects that engage communities in the design and construction of green spaces or eco-friendly buildings have been shown to increase individuals' sense of ownership and responsibility towards their environment [9]. Additionally, participatory design processes often result in the incorporation of local knowledge and cultural values into design solutions, leading to more contextually appropriate and

Fig. 15 Thin, translucent Rock-Skin-Casts illuminated. Photograph by Félix Michaud, 2022. Commissioned by and courtesy of Studio Pararaum

environmentally sensitive outcomes [10]. In the Rock Skin Pavilion, the participatory design team had similar experiences (see Fig. 22).

> Visiting the natural locations where initial casts were taken, designing with or alongside nature, and looking at the natural world as a source of curiosity and inspiration was a new approach for me.
>
> I felt most connected to nature when manipulating the latex casts of the rocks. Beyond being literally in nature and painting latex on rocks, I felt a sense of connection to nature when looking at the latex casts when degloved from the rocks. I think the latex casts to me acted as an index for the rocks, and found it fascinating…

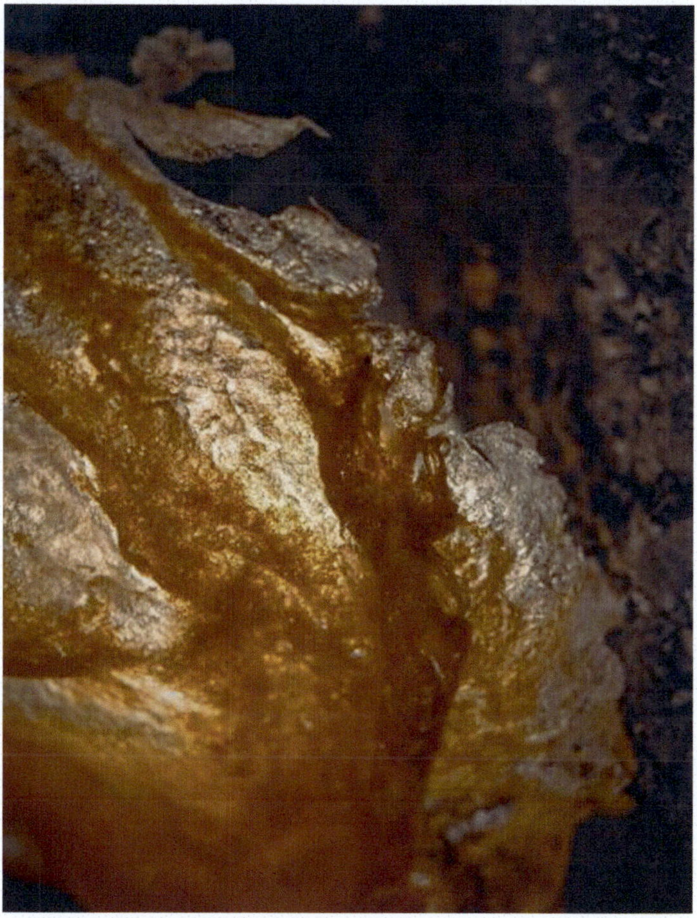

Fig. 16 Detailed view of the Rock-Skin-Cast, making visible the geological texture of the underground mine. Photograph by Félix Michaud, 2022. Commissioned by and courtesy of Studio Pararaum

> Iterating through the bioplastic colour tests using natural ingredients was definitely a memorable moment of my involvement in this project. So much research and prep work was involved in every step of the design process. This was a really fascinating slowing of pace in order to better understand the material and its properties that I really grew to cherish. It made me question how many short cuts and assumptions inform traditional design processes and pedagogies where the final product is deemed as the only thing that matters.

By involving communities in the design of their built environment, participatory design processes promote a sense of belonging, agency, and social cohesion, which are essential for positive mental health outcomes (see Fig. 23).

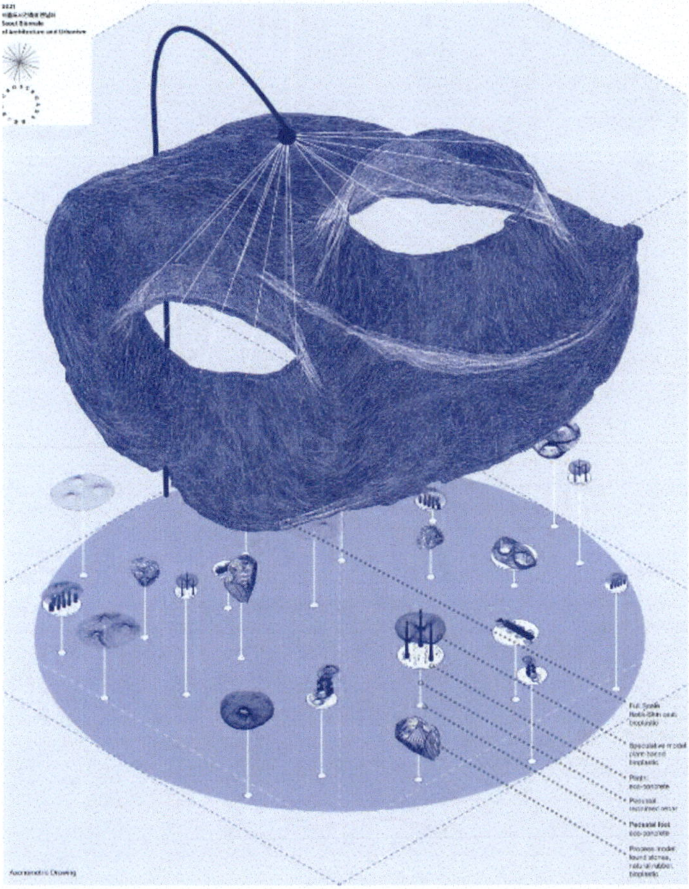

Fig. 17 Axonometric drawing of the Rock Skin Pavilion, submitted as part of the proposal for the SBAU open call. Drawing by Studio Pararaum, 2021

Participating in this project has heightened my intimacy with nature. As time progressed, I sought out valleys to find naturally shaped stones, and through the process of casting bioplastics onto these stones, I gained a deeper understanding of the essence of materials how it naturally bonds together due to properties itself. I still cannot forget the breeze of the wind, the smell of the grass, flowing sound of the valley, the colour of the green fields, the shadow of the tree casted on the rocky floor still left as a pleasant memory that I could experience during working time.

Fig. 18 Still from a process video showing the installation of the Rock Skin Pavilion at the Seoul Biennale of Architecture and Urbanism (SBAU), Seoul, South Korea. Video by Subin Kim of Studio Pararaum, 2021

> It has helped me understand how much my body innately craves contact with the natural environment. I never knew how powerful of a healer that nature can be on my psyche, but through this project, recognize that spending time in connecting to nature does help make me feel good. This project has also helped me consider the ways in which humans and our environment are interconnected, and that seemingly inert natural phenomena like soil, rocks and moss are in fact living beings that share our space.

Building on existing literature that links increase in CNS scores to behavioral changes, the design team also shared examples of ways in which participating in the Rock Skin design process had impacted their own day-to-day behavior around environmental stewardship and sustainable practices:

> I was able to better understand the gravity of a single individual's impact to extract from or negatively impact the natural environment, as well as see that public practices in the name of Environmental Sustainability does not adequately repair human damage or contact. In my daily life, I more frequently consider the natural public spaces that I enjoy and how my actions can possibly reduce or lengthen the lifespan of it.
>
> **I am more likely to take the time to care about small actions and process**
>
> I have made an eco-friendly plastic shopping bag at home using the same process of making the rock skin in order to approach the sustainable perception. Throughout this project, I have gained knowledge related to sustainability in materials which brought me a step forward to keeping nature safe.
>
> **I think working on a project that intrinsically demands a lot of reiterating, recycling, and, at times, waste, made me pay closer attention to the longevity and structural integrity of every material that exists around us.** It has also led me to question the ways

Table 1 Connectedness to nature scale questionnaire*

Please answer each of these questions in terms of the way you generally feel. There are no right or wrong answers
Using the following scale, in the space provided next to each question simply state as honestly and candidly as you can what you are presently experiencing
1 = Strongly Disagree, 2 = Disagree, 3 = Neutral, 4 = Agree, 5 = Strongly Agree

#	Question
1	I often feel a sense of oneness with the natural world around me
2	I think of the natural world as a community to which I belong
3	I recognize and appreciate the intelligence of other living organisms
4	I often feel disconnected from nature
5	When I think of my life, I imagine myself to be part of a larger cyclical process of living
6	I often feel a kinship with animals and plants
7	I feel as though I belong to the Earth as equally as it belongs to me
8	I have a deep understanding of how my actions affect the natural world
9	I often feel part of the web of life
10	I feel that all inhabitants of Earth, human, and nonhuman, share a common 'life force'
11	Like a tree can be part of a forest, I feel embedded within the broader natural world
12	When I think of my place on Earth, I consider myself to be a top member of a hierarchy that exists in nature
13	I often feel like I am only a small part of the natural world around me, and that I am no more important than the grass on the ground or the birds in the trees
14	My personal welfare is independent of the welfare of the natural world

* The content of the questionnaire is according to the study on CNS by Mayer, F.S. & McPherson Frantz, C. (2004). The connectedness to nature scale: A measure of individuals' feeling in community with nature. *Journal of Environmental Psychology, 24*, 503–515

in which we dispose of materials and objects that are deemed broken or damaged and how many people have subtly shifted away from mending and opted towards replacing. I also often question the quality of things that are being made today as we exist in a society that prioritizes convenience, quickness, low costs, and high profits. There is too much distance between people and the items they use temporarily and then discard of.

In conclusion, the participatory design approach of the SBAU Rock Skin pavilion project effectively bridged abstract sustainability attitudes with tangible actions, fostering ecological empathy and sustainable behaviors among participants. By involving individuals directly in the design and fabrication process, this approach not only enhanced their understanding of the material life cycle but also inspired lasting environmental stewardship in their daily lives (see Fig. 24).

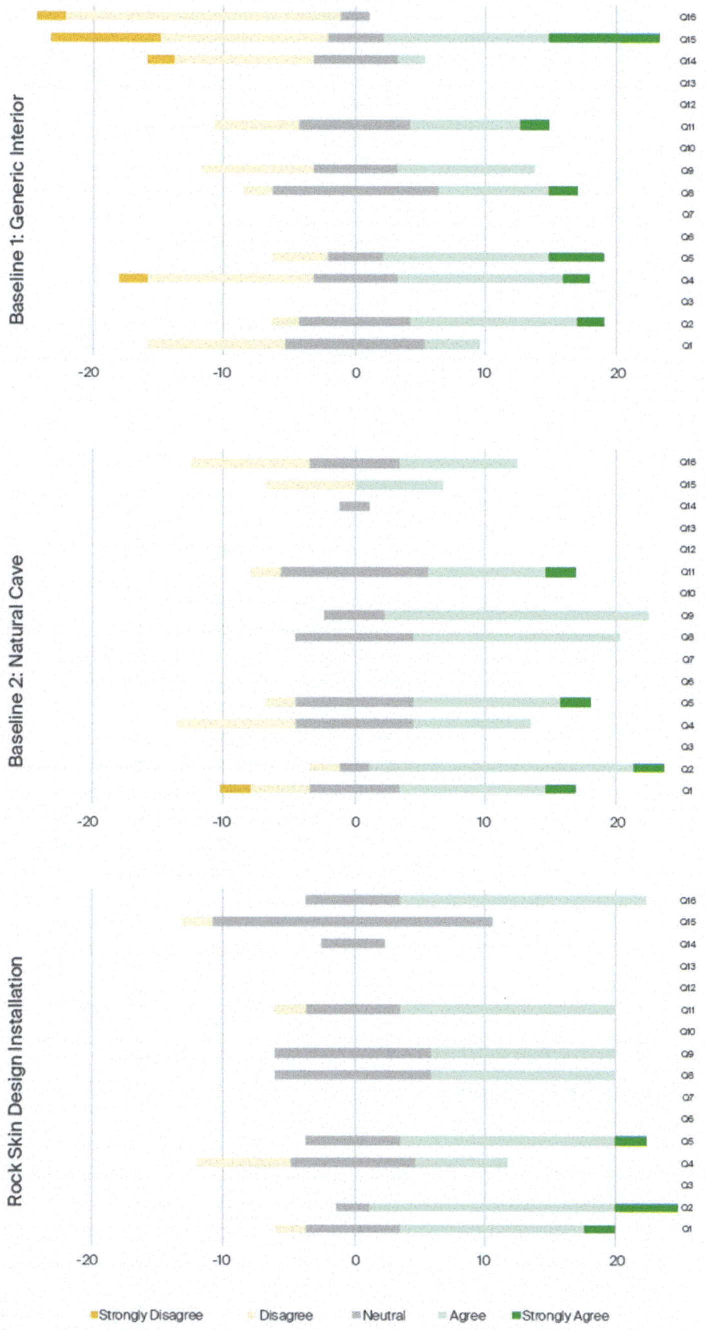

Fig. 19 Overview of the survey results from participants who visited the Rock Skin Installation, either physically or digitally, at the SBAU 2021, without any involvement in the design process

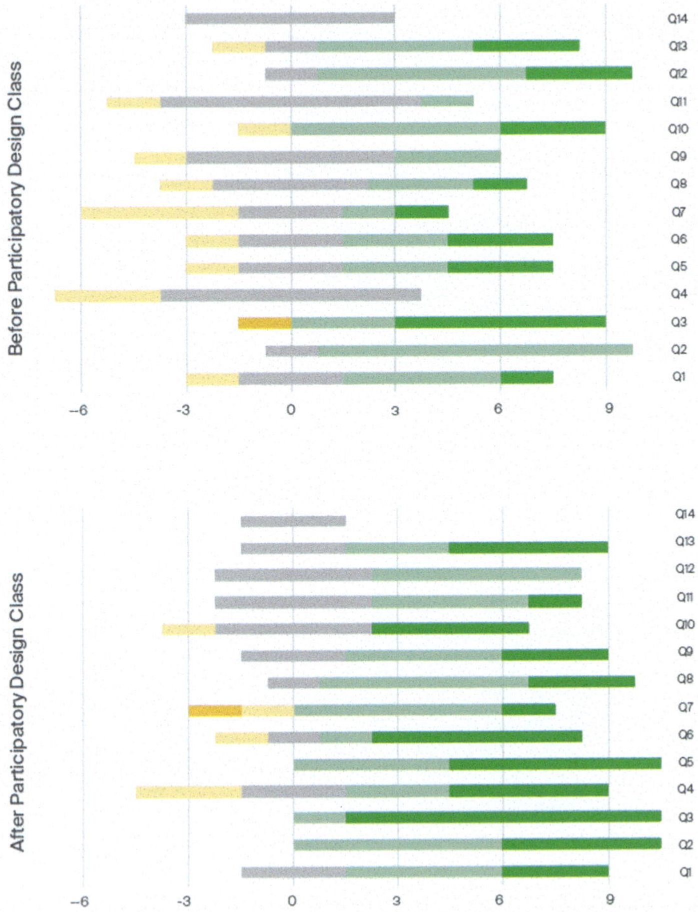

Fig. 20 Overview of the survey results from students who participated in the participatory design workshops in 2021

8 Future Recommendation: Participatory Design for Ecological Empathy

Our findings demonstrate the importance and potential of integrating participatory design principles into architectural education and practice. We recommend updating Arnstein's Ladder of Citizen Participation to center environmental stewardship to highlight and bolster the importance of community involvement in sustainable design. By 'climbing the rungs' of Arnstein's ladder of citizen participation in an environmental design context, power and knowledge can be redistributed from governments, architects, and planners to citizens, fostering a greater sense of community ownership and engagement.

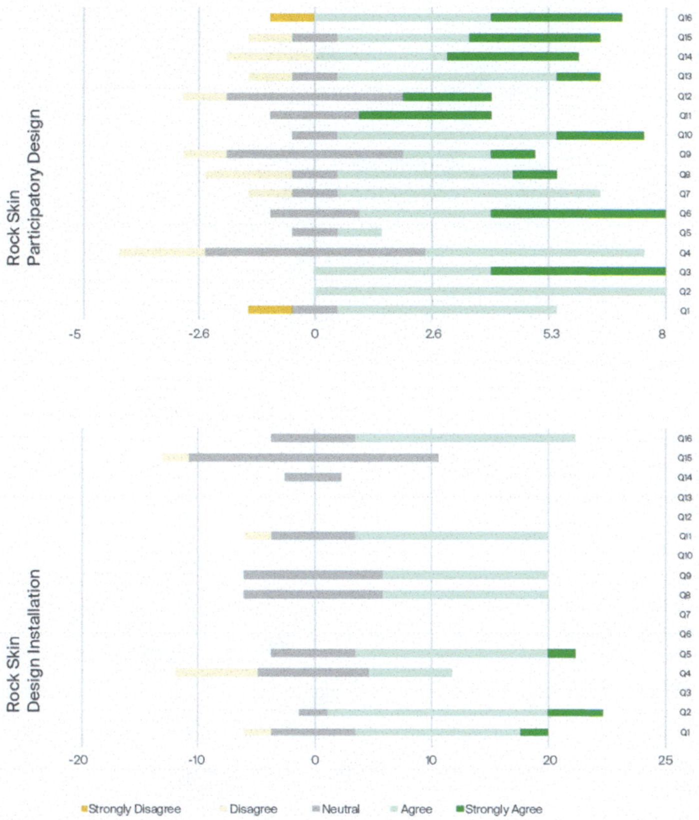

Fig. 21 Overview of the survey results from students who were involved in the research, design, and prototyping of the Rock Skin Pavilion at the SBAU 2021

Historically, citizen participation in Canada emerged in response to the amendments to the National Housing Act in 1949. Reaching its political peak in the late 1960s, guidelines for citizen participation were formalized in the Canadian legislature in 1969—the same year Arnstein's Citizen Participation Ladder was first published [5]. This parallels movements in the United States and the United Kingdom on similar timelines. These initiatives emphasized what true participation involved (citizens having the power to decide what happens in their own neighborhoods) and how this was different from other forms of tokenism (a chance to be heard, or an explanation of which policies were adopted and decisions were already made).

Against this historical context, it is essential to reimagine Arnstein's ladder [11]. Building on Zhang's (Studio Pararaum) previous work that remapped Arnstein's ladder to include design [12] (see Fig. 25), we recommend a future study of remapping the ladder to include environmental stewardship.

Fig. 22 Image showing the process of degloving a latex mold from a bioplastic cast. Experiment and photograph by Amy Yan, 2021

Fig. 23 **a** and **b** Process image of experiments in tinting bioplastic with natural dye colourants from foraged materials. Studio Pararaum experiments supported by design research team Georgia Barrington, Joelle Poitras, Johanna Weber, and Natalia Zuchniak. Photograph by Johanna Weber, Joelle Poitras, and Natalia Zuchniak. Copyright 2021 Studio Pararaum

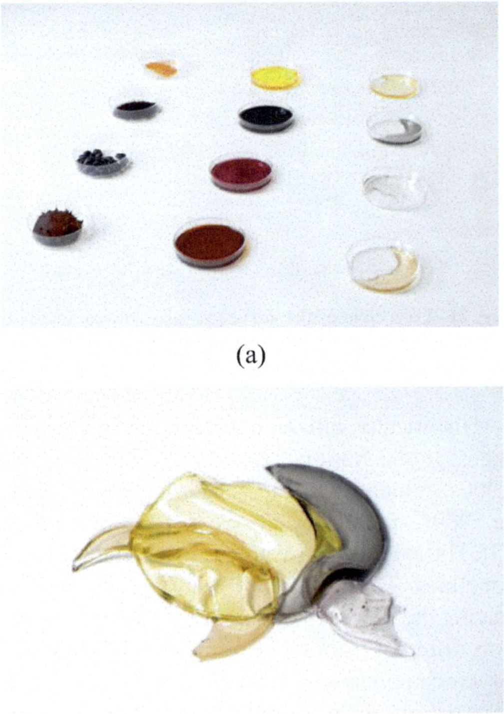

(a)

(b)

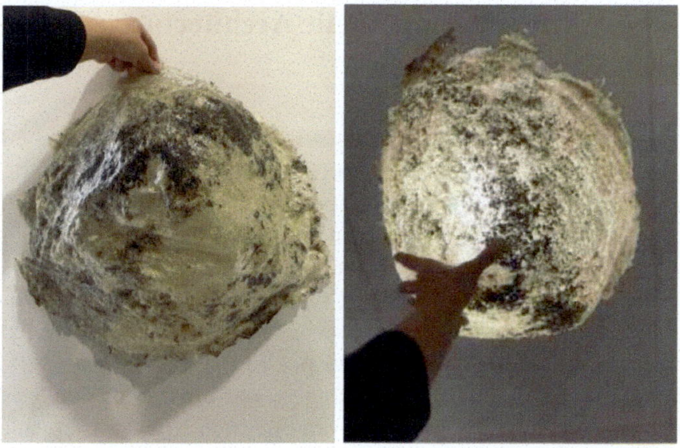

(a) (b)

Fig. 24 **a** and **b** Process images from the iterative prototyping of mid-scale Rock-Skin-Casts in bioplastics. Studio Pararaum prototyping and photograph by Georgia Barrington of Studio Pararaum, 2021

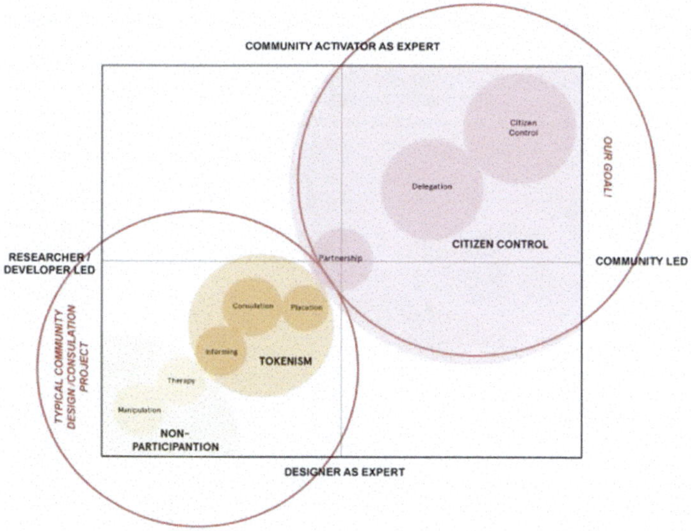

Fig. 25 Project Planting Imagination's Two-axis representation of Arnstein's Ladder of Citizen Participation as mapped against design and community expertise. From 'Planting Imagination: Community Co-Design for Chinatown Toronto I Impact Report,' by Linda Zhang, UWSpace, 2023

9 Conclusion: Toward Empathetic Architecture and Sustainable Futures

By embracing alternative material processes informed by the CNS, the Rock Skin Pavilion at SBAU positions itself not merely as a method of construction with sustainable materials, but challenges the norms of sustainable practice. With the novel approach of embedding geological memories in the materiality and involving inhabitants in its design and creation process, this intervention serves as a poignant reminder of the interconnectedness between what happens below and what emerges above, helping visitors comprehend the process of removal as an inevitable part of the experience of the city. Specifically, the Rock Skin project renders the extraction under the mountain visible while at the same time refusing to deplete the mountain. Through this process, we aim to elevate the collective consciousness around the consumption of resources and extraction processes implicit in the construction of our cities. It aims to act as a catalyst toward a sustainable planet by situating a design process more compassionately in between the underground and urban spaces, between our behavior and the climate challenges (see Figs. 26 and 27).

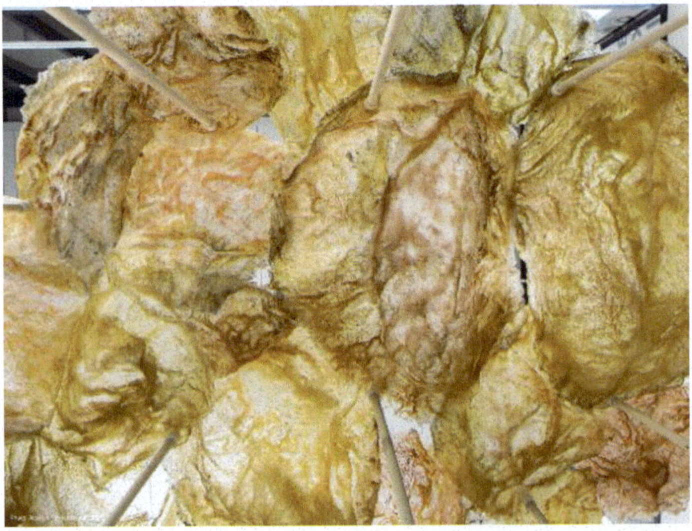

Fig. 26 Experience of being under the Rock Skin pavilion. Photograph by Jocelyn Reynolds, 2021. Commissioned by and courtesy of Studio Pararaum

Fig. 27 Rock Skin Pavilion prototype. Photograph by Jocelyn Reynolds, 2021. Commissioned by and courtesy of Studio Pararaum

Acknowledgements PRINCIPAL ARCHITECTS:

Meng Li + Linda Zhang (Studio Pararaum).

DESIGN TEAM:

Georgia Barrington (Project Lead), Reese-Joan Young and Amy Yan.

PROTOTYPING TEAM:

Natalia Zuchniak, Joelle Poitras and Johanna Weber.

SEOUL BIENNALE INSTALLATION TEAM:

Subin Kim 梯 熱綴 (Team Lead), Eunbin Cindy Hong, Lee Dongwon 檜 翕錳 + Sieun Lee 檜 衛擎.

ROCK-SKIN RESEARCH STUDIO STUDENTS:

Monica Beckett, John Vilmor Sicat, Amy Yan, Robin Grundy, Laura Trinchini + Eden Steenbakkers Lucuik.

FABRICATION SUPPORT:

Meimei Yang, Nora Alkeyat, Carol Kaifosh, Amanda Bly, Alexander Verni, Jonathon Anderson, Adrian Kenny, Buthaina Ahmed + Alda Escareno.

Funding Support:

Pro Helvetia Swiss Art Council, Ikea Foundation Switzerland, Toronto Metropolitan University University Internationalization at Home and SSHRC Explore Grant as well as the Seoul Biennale of Architecture and Urbanism.

IN-KIND SUPPORT:

Stiftung Gipsmuseum Gipsbergwerk Schleitheim, Stiftung Landesplattenberg Engi, Ontario Place, Design + Technology Lab at the Creative School + TMU SID 3D Studio Workshop.
PHOTOGRAPHER + VIDEOGRAPHY + DRONE PILOTS:
Stefan Müller, Félix Michaud, Jocelyn Reyolds, Meng Li + Linda Zhang (Studio Pararaum).

Competing Interests Relevant to the content of this chapter.

Ethics Approval This research study, *Design for Sustainable Behaviour: New Metrics for Architecture/Interior Design and User Behaviour,* has been reviewed and approved by the Toronto Metropolitan University Research Ethics Board (REB 2020-503) and is supported in part by funding from the Social Sciences and Humanities Research Council (TMU internal SSHRC Explore Grant). For additional information, please contact: Linda Zhang, Assistant Professor, University of Waterloo, School of Architecture linda.zhang@uwaterloo.ca.

References

1. Skin, Rock. "Rock-Skin." Studio Pararaum, www.rock-skin.com. Accessed 15 May 2024.
2. Mayer FS, McPherson Frantz C (2004) The connectedness to nature scale: a measure of individuals' feeling in community with nature. J Environ Psychol 24:503–515
3. Tonner J (2017) Nachhaltig nachhaltig?: Determinanten und Interventionen für kurz- und langfristig nachhaltiges Verhalten. Doctoral Thesis, University of Basel, Faculty of Psychology, pp 40–43. https://edoc.unibas.ch/56246/
4. Kouprie M, Visser FS (2009) A framework for empathy in design: stepping into and out of the user's life. J Eng Des 20(5):437–448. https://doi.org/10.1080/09544820902875033
5. Arnstein SR (1969) A ladder of citizen participation. J Am Inst Plann 35(4):216–224. https://doi.org/10.1080/01944366908977225
6. Gregg M, Seigworth GJ (2010) The affect theory reader. Duke University Press
7. Tonner J (2017) Nachhaltig nachhaltig?: Determinanten und Interventionen für kurz- und langfristig nachhaltiges Verhalten. Doctoral Thesis, University of Basel, Faculty of Psychology, p 98. https://edoc.unibas.ch/56246/
8. Stössel-Sittig I, Stehrenberger W, Stamm U (2005) Das Schleitheimer Gipsgewerbe: Geschichte, Geologie, Museum, Bergwerk. Stiftung zur Förderung des Gipsbergwerkes und Gipsmuseums Schleitheim
9. Puskás N, Abunnasr Y, Naalbandian S (2021) Assessing deeper levels of participation in nature-based solutions in urban landscapes: a literature review of real-world cases. Landsc Urban Plan 210(2):8. https://doi.org/10.1016/j.landurbplan.2021
10. Weymouth R, Hartz-Karp J (2019) Participation in planning and governance: closing the gap between satisfaction and expectation. Sustain Earth 2(5):10–11, 13. https://doi.org/10.1186/s42055-019-0012-y
11. Roberts A, Kelly G (2019) Remixing as praxis: Arnstein's ladder through the grassroots preservationist's lens. J Am Plann Assoc 85(3):301–320. https://doi.org/10.1080/01944363.2019.1622439
12. Zhang L (2023) Planting imagination: community co-design for Chinatown Toronto | Impact report. UWSpace. http://hdl.handle.net/10012/20104

Reimagining Subway Station Design: Beyond Prescriptive Approaches

Özgür Öztürk

Abstract This work expands the conventional understanding of subway stations by introducing innovative design strategies that prioritize spatial organization, crowd management, urban context integration, and multimodal transportation options. Traditional guidelines often neglect these critical aspects, focusing instead on layouts, materials, lighting, and capacity. Drawing on global examples that challenge these norms, this work advocates for flexible, context-specific approaches inspired by past architectural movements. By promoting creativity and adaptability from concept to construction, it provides a comprehensive road map for integrating urban planning, lighting design, and public health elements, ultimately transforming subway stations into vibrant, multifunctional spaces.

Keywords Subway station design · Underground design

1 Introduction

Do underground stations only serve as points of transition between underground and aboveground transportation networks, or do they possess a greater significance within the urban fabric? The creation of an underground space involves a complex design procedure that requires collaboration across several design teams.

Transportation systems, akin to other infrastructural endeavors, demand meticulous planning, organizational prowess, and adept utilization of innovative technology and construction methodologies. Their implementation on vast scales within urban landscapes, predominantly at the behest of governmental bodies, underscores the complexity inherent in their development. Such undertakings necessitate seamless collaboration across diverse engineering and design disciplines, highlighting

Ö. Öztürk (✉)
Faculty of Architecture, Hacettepe University, Ankara, Turkey
e-mail: ozgur.ozturk@hacettepe.edu.tr

© The Author(s), under exclusive license to Springer Nature Singapore Pte Ltd. 2025
M. Agrawal (ed.), *Underground Spaces for Climate Resilience and Sustainability*, Advances in 21st Century Human Settlements,
https://doi.org/10.1007/978-981-96-0547-7_14

their categorization as infrastructural projects overseen mostly by engineers. Regrettably, this emphasis often sidelines architectural and urban design considerations, overshadowed by structural and budgetary concerns.

This technical orientation engenders an environment where practicality takes precedence, often at the expense of contextual sensitivity. Consequently, transport stations and facilities often lack the architectural finesse and urban integration that could enrich their surroundings.

Amidst these challenges lies a fundamental question: are transport buildings merely integral components of infrastructure, or do they merit distinct recognition? Instead of relegating them to mere structures, it is perhaps more apt to envision them occupying a conceptual realm which goes beyond conventional classifications. In this view, they straddle the line between infrastructure and superstructure, embodying both physical functionality and intangible urban significance.

2 Traditional Subway Design Process

The architectural design process for underground stations is predominantly driven by technical concerns, where every aspect of design is meticulously guided by established standards and principles. A thorough examination of subway station design principles reveals a myriad of guidelines established by different countries and organizations. One can find an abundance of standards and guidelines from different countries functioning as subway and railway operators, disability acts, fire protection agencies, standard institutions, municipalities, and international cooperation agencies. Notably, while a sizeable portion of these guidelines addresses critical aspects such as fire safety, accessibility, and minimum design criteria for station components, there appears to be a dearth of specific guidance solely focused on station design itself. By using the parameters as design inputs and applying methods and calculations from these documents, it is possible to determine parameters such as the width of the platform, the area of the concourse, the width of the staircase, the number of turnstiles, the number of escalators, the number of ticket machines, the number of ticket issuing windows, and the necessary auxiliary spaces. In addition, designers also establish a comprehensive framework for organizing underground spaces, which consist of three distinct types of spaces: passenger spaces, technical spaces, and supplementary spaces. By using this approach, it is possible to structure a predetermined subterranean area easily (Fig. 1).

As a result, designers now have a predetermined blueprint that can be applied to any available area along the proposed route. In typical approaches, stations are characterized by their repetitive design where one can observe certain differences between stations in their entrance constructions, which vary according to their urban surroundings. Besides, these predetermined closed boxes mostly rely on artificial lighting and HVAC systems instead of any natural ventilation or illumination. Consequently, the prevailing recommendation for the landscape design of subway stations is often to restore the surface to its previous state, unless a comprehensive plan for integrating

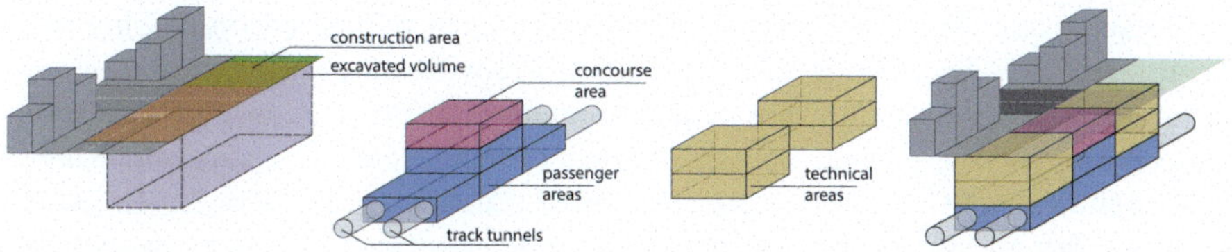

Fig. 1 Basic subway station components and layout

mass transit is devised. Moreover, this procedure ultimately produces standardized concrete pumps that move individuals across the cities.

The current linear process of a conventional process of subway station design can be seen in Fig. 2 which also displays the design inputs into three sections: Site Surveys, Transportation Surveys, and Operational Information.

Fig. 2 Existing design process of subway stations showing with the prescriptive methods

Site surveys collect data regarding potential transit routes, encompassing both the surface and subsurface levels. Map surveys provide data on the current usage, accessibility, coordinates, elevation, and orientation of urban areas at ground level. Geotechnical surveys provide data on the underground, seabed, or subsurface areas where significant development occurs in subway stations. These surveys offer essential data to designers regarding the physical characteristics, such as the soil and rock composition, of the earthwork which directly impact the reinforcement structures and structural design of the stations.

Transportation surveys are crucial investigations that reveal the patterns of mobility in cities, as well as identify potential areas for future growth and the optimal placements for subway stations. Designers obtain crucial data on passenger loads and peak hour passenger loads from these surveys, which directly impact the station design in the traditional approach.

Operational information provided by the line operator includes crucial data for shaping the technical features of the stations as well as information regarding the size and capacity of the vehicles, enabling the determination of numerous technical considerations and calculations. Vehicle dimensions play a crucial role in determining the length of the platform and the design of the platform separating doors (PSD), which are intended to divide the passenger area from the train tracks. Engineering inputs are generated to establish the fundamental aspects of the subway stations, in addition to the initial design inputs. This procedure involves the collaboration of transportation systems design engineers, geotechnical engineers, structural engineers, and geomatic engineers. They work together to produce comprehensive reports and documentation that are essential for the development of the subway line and its stops. During this stage, several tasks such as the creation of track alignment, geotechnical survey report, track tunnel, construction operation planning, retaining wall design, and foundation designs are carried out as well as the design of different types of stations and platforms. While engineers play a significant role in shaping these procedures, the station type and platform types are governed by architectural design processes.

As Kishii [1] highlights the design of a station extends beyond its project bounds. Nevertheless, the process of creating identical concrete structures and interring them across the subway route poses challenges and bears resemblances to flat-pack furniture, but on a larger architectural level. Similarly, Lang [2] argues against the modernist approach to design, which focuses on solving circulation and technical issues and creating a façade design but neglects to build a meaningful relationship between human needs, the environment, and their interrelation. Nevertheless, it can be enhanced by thoroughly examining each station's location, urban patterns, and users.

3 Elements of Subway Passenger Experience—And Ways of Improving It

Subway stations are designed to facilitate efficient passenger flow and enhance the overall experience. Pedestrian movement within stations typically follows a predefined path, except in underground stations with advanced guidance systems (Fig. 3). Typically, when passengers enter from entrance structures they move to concourse areas, where they access controlled areas and amenities like ticket vendors and shops, into controlled zones via turnstiles. Depending on the country and the context, security measures may be present at this stage. Inside the controlled area, passengers navigate through various circulation components like stairs and escalators toward platforms where signage aids the pedestrians as the main means of navigation in the underground setting. Finally, passengers reach platforms to board trains.

Through this linear pathway, pedestrians establish a decision-making process affected by factors such as either attracting or repelling their route choices (Fig. 4). Attracting factors, like social interactions or street markets could draw pedestrians toward certain paths, while repelling factors, such as low illuminated corners, social insecurity, or loitering, could deter them. Therefore, it is important to develop a design approach where these factors are considered both for functionality and pedestrian experience. However, it is essential to recognize that what may be appealing in one context can become repelling in another such as the variety of crowd, noise, and even smell. As a result, while a standard spatial layout may suffice, a more holistic

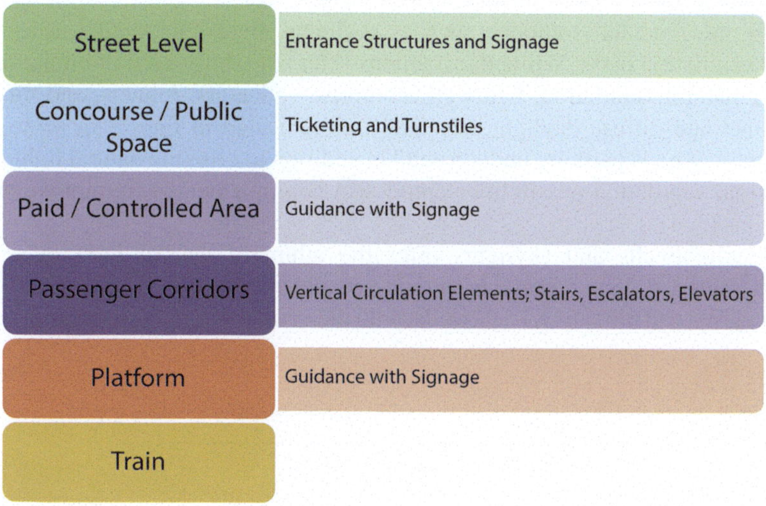

Fig. 3 The pedestrian flow outline in a subway station consists of sequential station zones arranged from the surface to the train, together with other elements and personnel assisting in the movement process

Fig. 4 Attracting and repelling factors in a pedestrian decision-making process to select the route

approach is advisable to create dynamic environments that challenge the conventional perception of stations as mere transitional spaces.

3.1 Challenges of Underground Systems

As the existing methodology suggests a linear design process where underground spaces are generated by guidelines, underground spaces turn into predetermined spaces where they are just decorated with various graphic and interior designs. In the context of underground, where being underground is associated with dark spaces with limited air quality, designers can challenge this notion by bringing natural elements such as light and ventilation inside the stations. By doing so, it is also possible to boost energy efficiency and psychological well-being in addition to the technical complexities.

There are potential solutions such as skylights, lightwells, wall washers, and natural ventilation ports. While the skylights offer straightforward illumination and possible natural ventilation, lightwells comprise collectors, funnels, and diffusers to channel and diffuse daylight effectively, as indicated in Fig. 5. In some cases where the station is partially underground in addition to bringing natural light inside the station, ventilation possibilities could also be implemented to improve station conditions [3, 4].

While lightwells and skylights offer the allure of natural light in subway stations, their installation presents intricate considerations covering lighting, structural integrity, and fire safety. Lightwells, with their complex structure, require careful

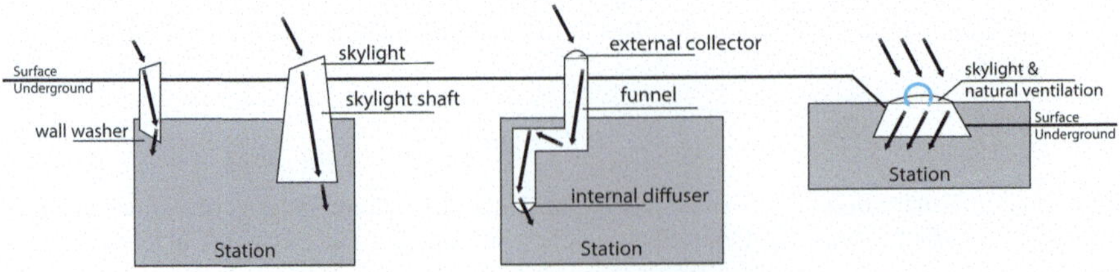

Fig. 5 Basic schemas of skylights and lightwells in subway stations

planning to balance lighting design elements and structural challenges, particularly in locations where direct access to the surface is not feasible. Despite their indirect delivery of daylight, they still pose challenges for fire protection and security, as they can act as ventilation shafts and are vulnerable to damage from vehicles and pedestrian activity.

Addressing these complexities is vital for underground structures, but the primary goal of maximizing daylight is to enhance pedestrian comfort and energy efficiency, not to replace artificial lighting. Designers can achieve this by integrating light wells or alternative openings with thoughtful landscape designs, thereby enhancing both the architectural appeal and psychological comfort of subway stations.

3.2 *Underground Station as an Urban Design Entity*

Considering the high volume of daily commuters in subway stations, it is pertinent to compare them with other structures catering to similar user traffic, like office buildings, hospitals, malls, or exhibition halls. Unlike these structures, which typically feature multiple pedestrian access points and various entrances for both private and service vehicles, subway stations often have a simpler design with fewer supplementary services nearby. While buildings like hospitals or sports facilities are evaluated based on their relevance and urban value, subway stations are often deemed acceptable as long as they feature prominent landmarks and wide doors to accommodate passenger flow. Admiraal and Cornaro [5] emphasize that subway stations should contribute to the urban fabric by making the underground visible from the surface, creating a cohesive "good space" that enhances the city. They advocate for a more progressive approach that integrates surface and subsurface areas rather than keeping them separate.

Conversely, when a subway station is located within an urban setting, it becomes a pivotal urban hub that significantly shapes the surrounding landscape through its design choices. Depending on the context, a station can serve to establish a new urban center, revitalize an existing one, or reshape the current environment.

While subway stations typically integrate with existing urban patterns and landscapes, there are also instances where infrastructural components are strategically placed in new areas to foster novel urban environments, as proposed by Allen [6]. This approach results in a diverse range of station designs, from individual stops to elaborate metropolitan hubs with interconnected subway lines.

However, it is also possible to trace different examples where stations not only restore the urban patterns but also reform them by incorporating various design elements, such as Toronto TTC Line 1 Extension Stations [7] or Yenikapı station proposals [8].

Unfortunately, due to the prevalence of prescriptive design methods and the prioritization of project management concerns like budgets and schedules, subway station designs often prioritize aesthetic enhancements determined by engineering requirements. However, the presence of advanced infrastructure projects in both underground and surface levels of subway stations presents a significant opportunity to reshape urban landscapes. Thus far, aside from modern instances, it has been challenging to find initiatives that fully exploit the potential of underground stations as urban entities.

However, it would be misguided to create subway stations that are disconnected from their urban surroundings and users. On the contrary, the conventional perception of subway stations should evolve to view them as a process for developing underground urban spaces that offer mobility options rather than merely constructing prescribed concrete structures serving different areas of the city. By adopting this approach, we can move beyond a reductionist mindset that solely focuses on functionality or individual components.

3.3 Establishing a Connection Between the Surface Level and Underground

The surface level of an underground station holds paramount importance as it must seamlessly integrate with the surrounding urban context. Design regulations often advocate for minimizing land disruption during construction, with construction companies typically held responsible for restoring the site to its original condition [9, 10]. However, there are alternative suggestions worth exploring beyond mere restoration, which could enhance the surface level in ways that complement the urban environment.

Typically, the standard approach for addressing the surface of subway stations involves restoring it to its original state upon completion of construction. This method maintains a distinct separation between the upper and lower levels, with signposts guiding pedestrians to the station's location on the surface. This architectural strategy is often chosen when the surface area of the station is integral to the surrounding city, such as when located beneath a significant urban feature like a historic square in need of restoration or beneath a street, highway, railway, or airport. Employing this approach helps to reestablish the city's previous connections. The repaired surface space can potentially be repurposed as a park, road, or local square, as depicted in Fig. 6. However, it is important to note that subway stations and their surfaces remain disconnected, except for the entrance constructions situated around the stations.

Challenging the traditional design approach opens the door to creating new landscapes at ground level around subway stations. Instead of merely restoring previous conditions, designers can craft diverse landscapes with varying elevations to cater to users' specific needs. With the introduction of new transit options facilitated by these stations, there is an opportunity to not just restore but enhance or transform public spaces. Designers can leverage excavated soil from station construction to shape

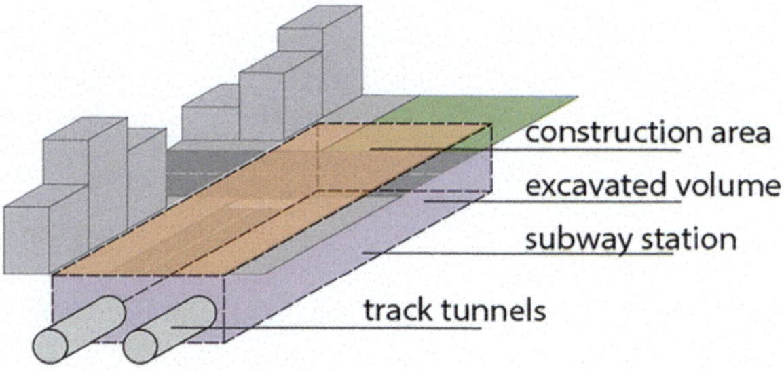

Fig. 6 Schematic view of a typical station

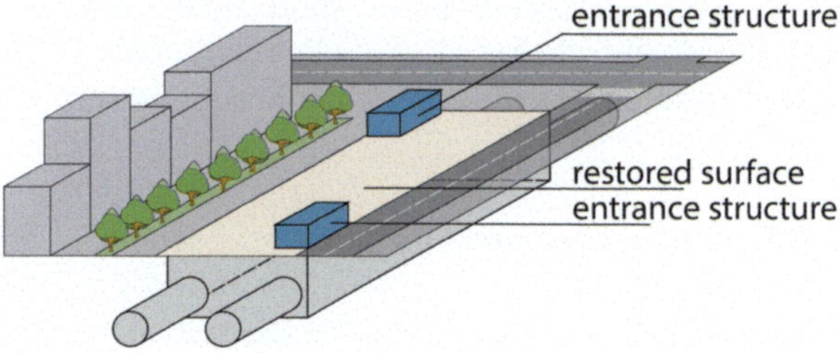

Fig. 7 A schematic representation of a station with numerous entrances, which are outlined in blue, situated at the surface level

innovative landscapes, aligning with sustainability standards for managing excavated materials. Successful execution could result in the transformation of station areas into vibrant public spaces offering a range of activities. Moreover, these landscape changes can also facilitate natural lighting and airflow, further enhancing the overall station experience. An example of this can be observed at the York University station in Toronto [11].

In many urban settings, there is potential to enhance subway stations and their surroundings with additional features. Given the dense population, stations often require multiple entrances equipped with stairs, escalators, and elevators for accessibility, which may be located separately or adjacent to the station, depending on surface conditions (Fig. 7). Entrances can also be adorned with canopies for shelter and as a distinctive symbol. However, in urban settings, these entrances may require more complexity, offering features like passenger-operated machines, shopping facilities, cash machines, tourism kiosks, waiting areas, and other transportation options. Similarly, the surface area of the station can be improved by incorporating new functions alongside redesign efforts. These approaches typically fall into three categories: open-air solutions, expanding the station as a superstructure, and hybrid solutions.

3.3.1 Open-Air Approach

Open-air solutions entail creating public spaces on surface levels, depending on design decisions. These areas can be outfitted with canopies and temporary structures to host various events (Fig. 8). While the conventional approach involves placing bus stations and transportation facilities in this vicinity, additional amenities like

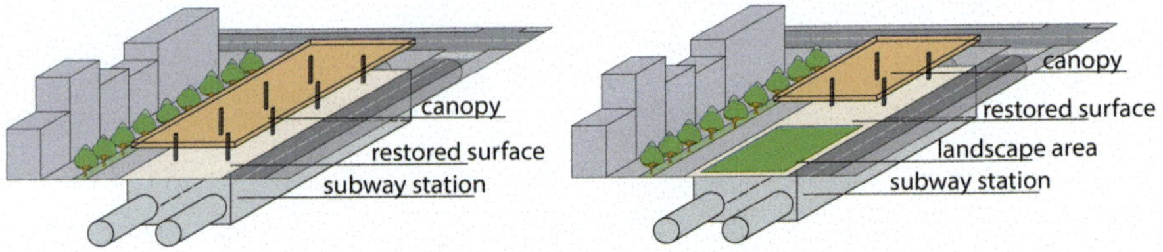

Fig. 8 Open-air surface solutions that could contribute to the improvement of the urban environment

marketplaces, gathering spaces, sports fields, and communal areas can be integrated. By incorporating these spaces above newly constructed underground areas, designers enhance station functionality while benefiting surrounding communities.

3.3.2 Superstructure Approach

Designers can also opt to expand the station by constructing a superstructure on the surface rather than adopting open-air solutions. However, this approach introduces intricate design considerations to accommodate numerous functions within the proposed structure (Fig. 9). Various components of the design must comply with safety and operational regulations. This method is suitable for densely populated areas where land is scarce or when designers aim to create a central focal point by incorporating multiple activities in the station area. Such structures can house government institutions, offices, public services, shopping areas, meeting spaces, houses of worship, and even hotels, establishing a self-contained economic ecosystem with a significant impact on its surroundings. Nevertheless, coordinating structural designs for both the superstructure and the subway station may pose challenges for designers.

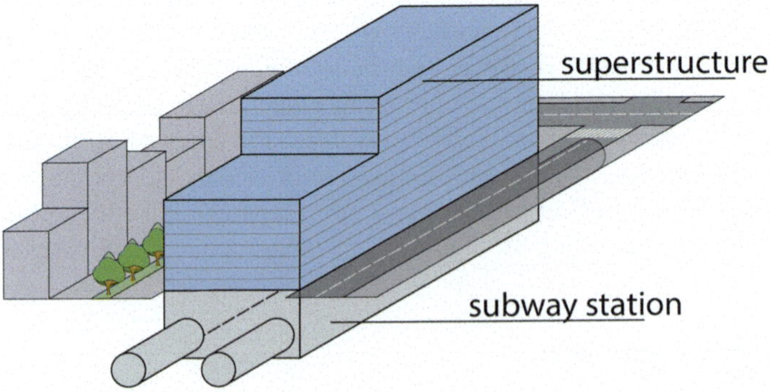

Fig. 9 A Superstructure solution that could utilize the surface to modify the urban environment

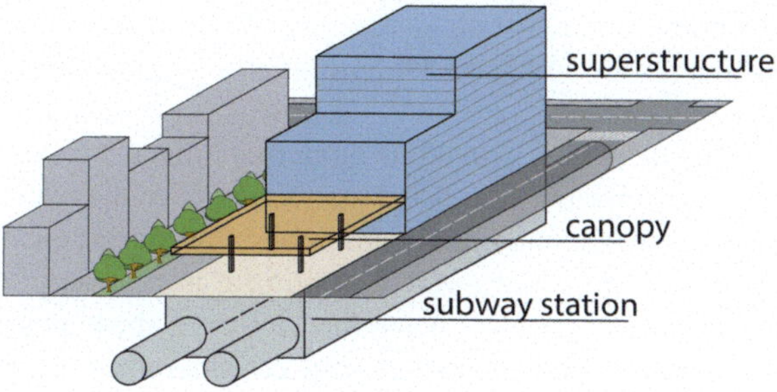

Fig. 10 A Hybrid solution that allows designers to integrate open-air solutions and superstructures for their surface design

3.3.3 Hybrid Approach

It is possible to integrate open-air and superstructure approaches to create a hybrid solution to address various design problems. Despite the overall complexity of this approach, it can improve the station and its urban surroundings. The designers have the choice of either integrating their work into the existing urban context or constructing its own components in order to generate new urban patterns. Yet, it is also important to note that a subway station is also an urban entity, not only in its affordable and reliable transportation services to its users but also in its potential urban center prospects. It is possible to state that stations could turn into urban centers for various communities [12]. Even though this is a difficult method in general, it is possible to recommend structures that might enhance the current situation by producing a concise urban assessment of the region. These buildings could include public service buildings, commercial areas, sports halls, parks, and recreational facilities, as well as healthcare facilities, and public education centers (Fig. 10).

3.4 Integrating Multimodal Transportation and Micromobility Around Subway Stations

The optimal range for accessing subway stations is typically considered to be around 1–1.5 kilometers in level reaching distance unless various transportation options are integrated into the stations [13]. These multimodal capabilities, such as park-and-ride, kiss-and-ride, and bicycle store-and-ride facilities, enable passengers from varying distances to easily access and utilize the subway network, enhancing the urban significance of the stations. However, integrating these options requires specific infrastructural and architectural arrangements, and grouping similar tasks and providing necessary infrastructure and utilities offers benefits in terms of cost, space, time, and resources.

Alternatively, various levels of mobility choices, including ride-sharing systems like bike-sharing, carsharing, scooter sharing, and e-bike sharing, have been shown to enhance pedestrian accessibility in densely populated metropolitan areas and offer quick and efficient transportation [14, 15]. Studies have demonstrated decreases in car purchases and personal automobile usage among users of bike-sharing systems [16].

Therefore, it is crucial to implement diverse transportation activities not only to enhance subway stations but also to improve the overall urban environment. Micromobility, when integrated with public transportation systems, is considered an alternative and potential remedy for traffic issues, despite debates about accessibility, safety, speed, and user behavior. However, these possibilities have sparked debates among users and require careful consideration. Integration should be part of the architectural design from the outset, with docking and charging stations for electric-powered scooters or bicycle facilities with supplementary services included in early design phases to boost accessibility.

3.5 *Health and Hygiene*

Subway stations require efficient planning not only for vehicle integration but also for public health considerations, especially in preventing the spread of diseases due to the substantial number of people passing through them daily. Establishing strict hygiene standards, including regular disinfection of surfaces, and providing hand sanitizing stations, is crucial. Optimizing ventilation systems is also important to reduce the spread of germs.

During the initial phase of the COVID-19 pandemic, there was a decrease in the use of mass transportation due to concerns about maintaining physical distance. Alternative solutions, such as bicycles and shared mobility systems saw increased demand [17]. However, transitioning from public to private transportation is not a viable long-term solution for urban planning and climate change considerations, as it would exacerbate congestion and environmental impacts [18, 19].

Reports from countries like France and Japan found no viral clusters associated with transportation networks, indicating that public transit systems did not contribute significantly to the spread of COVID-19. Nevertheless, it is important to address concerns about ventilation, sunlight, and minimizing time spent indoors to mitigate risks associated with future pandemics. Utilizing natural ventilation options, such as skylights and ventilation shafts, can help improve air quality within subway stations. However, challenges such as separating these structures from pedestrian and vehicle traffic and ensuring uncontaminated air intake need to be addressed in addition to their fire safety concerns during emergencies.

Well-organized stations that facilitate efficient movement and incorporate touchless fare collections and temperature control checkpoints can enhance safety during pandemics. Overall, lessons learned from the COVID-19 pandemic underscore the importance of developing alternative suggestions for closed spaces, particularly in terms of ventilation and lighting, to ensure public health and safety.

4 Discussion

Underground stations are integral parts of metropolitan life, serving millions of people daily. However, the common perception of station design primarily revolves around engineering considerations rather than architectural concerns. This chapter proposes an alternative approach where architects and engineers collaborate concurrently in the design process.

In this proposed design process (Fig. 11), the architectural design takes precedence over technical considerations, such as station and platform typology, which traditionally dictate space organization. The initial phase involves conducting site surveys and determining station sites to gather data about the surrounding urban environments. This data informs the design of both passenger spaces within the station and the surrounding urban landscape, with a focus on creating interconnected areas that merge seamlessly with technical station requirements.

Traditionally, station design focuses on technical requirements and space allocations based on transportation needs, often resulting in monotonous and temporary spaces. However, in the proposed process, pedestrians are at the center of the design, and the station architecture serves to facilitate connections and activities while enhancing the urban environment. Scrutinizing established standards and rules allows designers to challenge conventional design approaches and create more vibrant and functional station areas.

To improve the current design process, it is essential to question the uniform approach that treats stations as mere transportation hubs. Design solutions should consider the station's role in the urban context and prioritize user experience, including clear signage and intuitive navigation systems. By integrating design elements that enhance usability and contribute to the station's identity, we can create spaces that are both functional and engaging, avoiding the perception of stations as nondescript or anonymous "non-places."

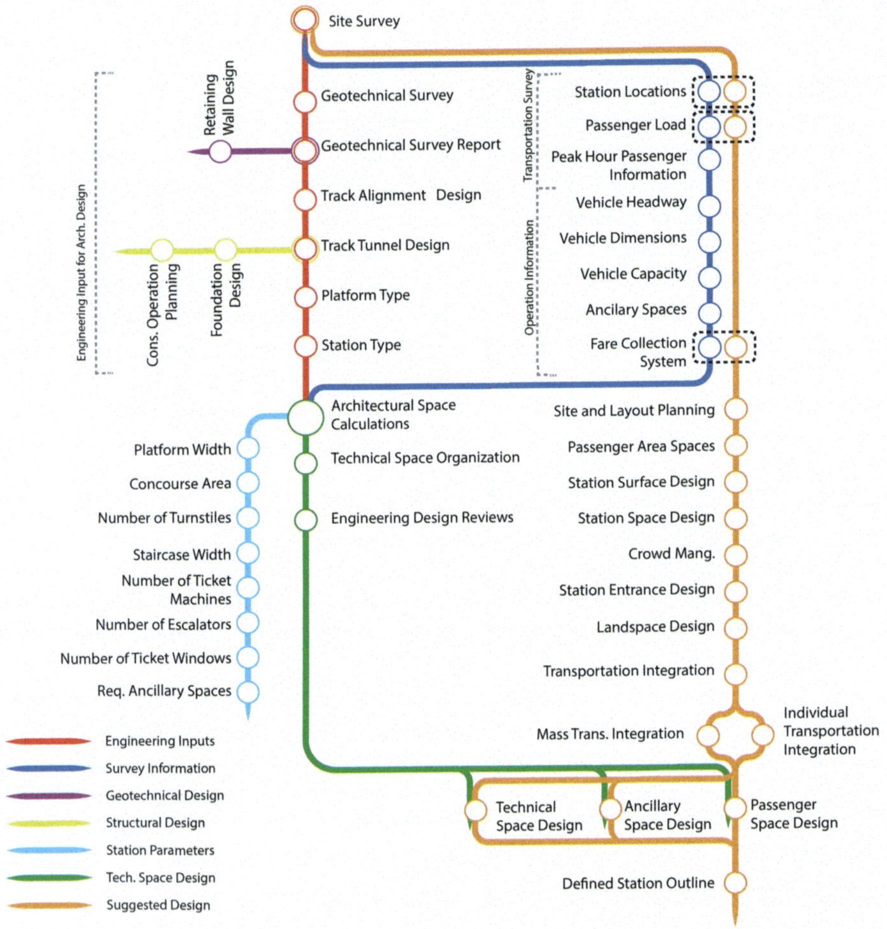

Fig. 11 Suggested subway station design workflow showing an alternative for the existing prescriptive design approach

Competing Interests This chapter is based on the author's Ph.D. dissertation "A Research on the Design Criteria of Subway Stations" submitted to the Graduate School of Natural and Applied Sciences of Middle East Technical University, Department of Architecture under the supervision of Prof. Dr. Soofia Tahira Elias Ozkan.

The authors have no conflicts of interest to declare that are relevant to the content of this chapter.

References

1. Kishii T (2016) Utilization of underground space in Japan. Tunn Undergr Sp Technol 55:320–323. http://www.sciencedirect.com/science/article/pii/S0886779815303515

2. Lang J (1994) Urban design: the American experience. Van Nostrand Reinhold, New York, NY, p 509
3. Molteni SC, Courret G, Paule B, Michel L, Scartezzini JL (2001) Design of anidolic zenithal lightguides for daylighting of underground spaces. Sol Energy 69:117–129. http://www.sciencedirect.com/science/article/pii/S0038092X01000652
4. Bouchet B, Fontoynont M (1996) Day-lighting of underground spaces: design rules. Energy Build 23(3):293–298. http://www.sciencedirect.com/science/article/pii/037877889500954X
5. Admiraal H, Cornaro A (2017) The impact of urban planning on the design and operation of stations and interchange hubs. In: 2017 STUVA research + practice, underground transportation and underground construction. Stuttgart: STUVA e.V., p 269–274
6. Allen S (1997) From object to field. AD (Architectural Des.) (5–6):24–31
7. Toronto Transit Commission (TTC) (2009) Toronto-York Spadina Subway Extension project steeles west station. Approval of Conceptual Design. Toronto
8. Yazman D (2011) Yenikapı Transfer Noktası ve Arkeopark Alanı Uluslararası Mimari Avan Projesi. Davetli Hizmet Alımı - Arkitera. https://www.arkitera.com/yarisma/yenikapi-transfer-noktasi-ve-arkeopark-alani-uluslararasi-mimari-avan-projesi-davetli-hizmet-alimi/. Accessed 30 Nov 2019
9. EGO Ankara Department of Transportation Planning and Rail System (1992) Ankaray Ankara light rail system, vol 1. Transportation Survey. Municipality of Greater Ankara, Ankara
10. JICA (2010) Preliminary Design (Phase 1). In: JICA preparatory survey on greater Cairo metro line No4 in the Arab Republic of Egypt, p 1–461
11. Toronto Transit Commission (TTC) (2009) Toronto-York Spadina Subway Extension project highway 407 station. Approval of Conceptual Design. Toronto
12. Öztürk Ö (2023) Evaluation of subway stations from an architectural point of view. In: Expanding underground—knowledge and passion to make a positive impact on the world. CRC Press, London, pp 829–834. https://www.taylorfrancis.com/chapters/oa-edit/10.1201/9781003348030-100/evaluation-subway-stations-architectural-point-view-öztürk. Accessed 2023 May 26
13. Griffin KW (2004) Station types and configurations. In: Building type basics for transit facilities. Wiley, Hoboken, New Jersey, pp 63–80. (Building type basics series)
14. Moran ME, Laa B, Emberger G (2020) Six scooter operators, six maps: spatial coverage and regulation of micromobility in Vienna, Austria. Case Stud Transp Policy 8(2):658–671. http://www.sciencedirect.com/science/article/pii/S2213624X20300171
15. Lazarus J, Pourquier JC, Feng F, Hammel H, Shaheen S (2020) Micromobility evolution and expansion: understanding how docked and dockless bikesharing models complement and compete—a case study of San Francisco. J Transp Geogr 84:102620. http://www.sciencedirect.com/science/article/pii/S0966692319302510
16. Shaheen S, Cohen A, Chan N, Bansal A (2020) Chapter 13—Sharing strategies: carsharing, shared micromobility (bikesharing and scooter sharing), transportation network companies, microtransit, and other innovative mobility modes. In: Deakin Land use, and environmental planning EBTT. Elsevier, pp 237–262. http://www.sciencedirect.com/science/article/pii/B978012815167900013X
17. Landis-Hanley U. The Guardian (2020) "Bicycles are the new toilet paper": bike sales boom as coronavirus lockdown residents crave exercise. https://www.theguardian.com/lifeandstyle/2020/apr/22/bicycles-are-the-new-toilet-paper-bike-sales-boom-as-coronavirus-lockdown-residents-crave-exercise. Accessed 3 Jun 2020
18. London City Hall. Mayor of London (2020) Mayor's bold new Streetspace plan will overhaul London's streets. https://www.london.gov.uk/press-releases/mayoral/mayors-bold-plan-will-overhaul-capitals-streets. Accessed 29 Jun 2020
19. Bliss L, Lin JCF, Patino M, Bloomberg CityLab (2020) Global data reveals inequality of pandemic travel. https://www.bloomberg.com/graphics/2020-coronavirus-transportation-data-cities-traffic-mobility/?cmpid=BBD061920_CITYLABMP&utm_medium=email&utm_source=newsletter&utm_term=200619&utm_campaign=citylabmostpop. Accessed 20 Jun 2020

Conclusion

Digging Deeper: The Future of Underground Urban Exploration

Mahak Agrawal

Keywords Urban underground space · Urban innovation · Urban climate resilience · Carbon management · Sustainability

As we draw the curtains on this subterranean adventure, it's time to face the facts: the underground is no longer just the domain of moles, earthworms, and mysterious tunnel-dwelling creatures from childhood tales. It's a space that's not just teeming with life—albeit of a somewhat spidery variety—but with potential that, if we dare to dig a little deeper, could revolutionise the way we live, work, and play in our cities.

Throughout our exploration, we've uncovered how the underground, once regarded as a last resort for infrastructure and a hiding place for secret lairs, can now stand at the forefront of urban innovation. This isn't just about finding new places to put things when the surface gets too crowded, though that's a nice bonus. It's about recognising that the ground beneath our feet is a critical resource—one that can help us tackle some of the most pressing issues of our time, from climate change and urban sprawl to social inequity and resource scarcity.

M. Agrawal (✉)
Steering Board Member (Urban Adaptation), ITA Committee on Underground Spaces (ITACUS), Geneva, Switzerland
e-mail: ma4100@columbia.edu

© The Author(s), under exclusive license to Springer Nature Singapore Pte Ltd. 2025

M. Agrawal (ed.), *Underground Spaces for Climate Resilience and Sustainability*, Advances in 21st Century Human Settlements,
https://doi.org/10.1007/978-981-96-0547-7_15

1 Reimagining Urban Space: A Third Dimension

If there's one thing this journey has shown us, it's that we've been thinking about space all wrong. For too long, urban development has been a two-dimensional affair, focused on expanding horizontally and, in more recent times, vertically. But in a world where surface space is becoming scarcer and more expensive than ever, it's time we started thinking in three dimensions.

The underground offers a rich, largely untapped dimension that can relieve the pressure on our overstretched surface. But we're not just talking about parking garages and subway systems here. Imagine a city where underground spaces are vibrant, multifunctional areas—places where people can live, work, learn, and even unwind. Think of underground neighbourhoods that are cool in the summer, warm in the winter, and unaffected by the whims of weather on the surface. Think of parks where the sun might not shine, but where innovative lighting and lush vegetation create a serene escape from the hustle and bustle above.

Yet, reimagining urban space is not merely a matter of digging holes and hoping for the best. It's about sophisticated planning, understanding the geology, hydrogeology, and even the acoustics of these spaces. After all, nobody wants to live in a flat where every footstep sounds like a drumbeat, or where the neighbour's karaoke sessions sound like they're happening right in your living room. Advanced engineering and design, supported by cutting-edge technologies like Building Information Modelling (BIM) and Geographic Information Systems (GIS), are critical to making underground spaces not only functional but also comfortable and enjoyable. The underground is, quite literally, a world of opportunity waiting to be unearthed.

2 Digging Deeper into Possibilities: The Role of Innovation

Innovation, in this context, is our shovel, our excavator, and our tunnel-boring machine—digging deeper into the earth and into the possibilities that lie within it. The examples we've explored throughout this book—whether it's the subterranean data centres of the Arctic or the underground farms of Tokyo—are a testament to the fact that with enough ingenuity, the challenges of the underground can be turned into opportunities.

But let's not kid ourselves: the underground isn't without its quirks. It's dark, it's damp, and it's not exactly the easiest place to install a decent Wi-Fi signal. Yet, these challenges are precisely where innovation thrives. We've seen architects and engineers turn former nuclear bunkers into chic co-working spaces, transform abandoned mines into energy storage facilities, and reimagine sewers as venues for avant-garde art installations. With each challenge comes a creative solution that not only overcomes the limitations of the underground but also enhances its unique character.

However, let's not forget that innovation isn't just about fancy gadgets and clever designs. It's also about changing the way we think about space. We need to shift our perspective from seeing the underground as a last resort to viewing it as a first choice. This requires breaking down the barriers between disciplines—getting geologists to talk to architects, sociologists to collaborate with engineers, and policymakers to listen to all of the above. It's about creating a new narrative where the underground is not an afterthought, but a cornerstone of urban planning (see Fig. 1).

Fig. 1 Innovation for underground spaces is driven by shovels, excavators, and tunnel-boring machines. Image by Mahak Agrawal

3 The Future of Urban (Underground) Development

As we peer into the future, it's clear that the underground will play a central role in shaping the cities of tomorrow. The pressures of urbanisation, climate change, and resource depletion mean that we can no longer afford to ignore the potential beneath our feet. The underground is not just a solution to these challenges; it's an opportunity to rethink how we build and live in our cities.

Consider this: as climate change continues to push the limits of our urban environments, the underground offers a climate-controlled haven, naturally insulated and less prone to the extremes of weather that are becoming increasingly common on the surface. Underground spaces can serve as refuges during heatwaves, storm shelters during hurricanes, and stable environments in regions prone to earthquakes or other natural disasters. In this context, the underground isn't just a convenience; it's a lifeline.

Moreover, the underground presents a chance to address some of the most stubborn issues in urban planning—like affordable housing, sustainable infrastructure, and social equity. By building downwards, we can create new housing stock without displacing communities or encroaching on green spaces. We can construct infrastructure that's both resilient and sustainable, powered by geothermal energy and integrated with above-ground ecosystems. We can design spaces that are accessible and beneficial to all, ensuring that the benefits of urban development are shared by everyone, not just those who can afford prime real estate on the surface.

But let's be real: none of this is going to be easy. Developing underground spaces requires significant investment, careful planning, and a willingness to navigate the complex regulatory and logistical challenges that come with building beneath the surface. It also requires a commitment to inclusivity, ensuring that these spaces are accessible to all members of society, not just a privileged few. But as we've seen throughout this book, the rewards are well worth the effort. The underground is not just the final frontier of urban development—it's the foundation upon which the cities of the future will be built.

4 Final Thoughts: The Ground Beneath Our Feet

In closing, it's worth taking a moment to reflect on the ground beneath us. For centuries, the underground has been a place of mystery and intrigue—a realm of myths, legends, and the occasional archaeological dig. But as we've seen, it's also a space of immense potential, waiting to be explored, developed, and integrated into the fabric of our cities.

So, as you walk the streets of your city, take a moment to think about what lies below. Imagine the bustling networks of transport, energy, and life that could be thriving just a few metres beneath your feet. The underground is no longer a place to be feared or ignored—it's a space of opportunity, innovation, and, dare I say, a bit of fun. After all, who wouldn't want to live in a world where your morning commute takes you through an underground park, your office is a stylish subterranean space, and your evening ends with a visit to an underground concert that rocks?

The future is beneath our feet, and it's time we started digging. Thank you for joining me on this journey into the depths. I hope it has inspired you to look at the world in a new way, to see the underground not just as a relic of the past, but as a vibrant, essential part of our future.

The manufacturer's authorised representative in the EU is Springer Nature Customer Service Centre GmbH, Europaplatz 3, 69115 Heidelberg, Germany. If you have any concerns regarding our products, please contact ProductSafety@springernature.com

Printed and bound by CPI Group (UK) Ltd, Croydon, CR0 4YY
01/10/2026
02225939-0001